Naturalists' Handbooks 30

Snails on rocky sea shores

JOHN CROTHERS

Pelagic Publishing
www.pelagicpublishing.com

Published by **Pelagic Publishing**
www.pelagicpublishing.com
PO Box 725, Exeter, EX1 9QU

Snails on rocky sea shores
Naturalists' Handbooks 30

ISBN 978-1-907807-15-2

Series Editors
S.A. Corbet and R.H.L. Disney

British Library Cataloguing in Publication Data
A catalogue record for this book is available from the British Library.

Printed and bound in India by Replika Press Pvt. Ltd.

Contents

Cover photographs: Peninnis Head, St Mary's, Isles of Scilly;
part of a breeding aggregation of dog-whelks (*Nucella lapillus*)
and their yellow egg capsules; two common limpets
(*Patella vulgata*) at home

Preface

Naturalists' Handbooks are written to assist people, of all ages and range of experience, to investigate the natural history around them. Many people first attempt to do this as a project undertaken as part of a school or university course, but that restricted time frame severely limits the type of investigation that can be contemplated. Naturalists' Handbooks are not so constrained, being written with the amateur naturalist also in mind.

Whilst not, at first sight, the most exciting component of the fauna, the snails found living on rocky sea shores must be amongst the most rewarding invertebrate animals to study. With a little practice, the species are easy to find, capture, identify, measure and mark. They don't bite, sting or run away when humans appear; moreover they are also tough and appear to survive handling without ill effect. Most species are present on the shore throughout the year and individuals of most live for several years. This author is firmly of the opinion that field investigations into the lives of common species will be the most rewarding; limpets, dog-whelks, topshells and winkles will offer a wide range of possibilities.

Acknowledgements

Thanks are due to the Field Studies Council, the Linnean Society of London and the Malacological Society of London for permission to reproduce the key illustrations, and to the Field Studies Council for permission to reproduce other material that originally appeared in the journal *Field Studies*. Thanks also to Mr P. S. Croft for fig. 48, Dr M. A. Kendall for fig. 31 and Dr. P. J. Hayward for K44.

The material presented in this book was accumulated, over many years, for the purpose of running field courses for school and university students. I am grateful to all those students who collected data for me and discussed their results; discussion with other staff (both those employed by FSC and those visiting the field centres with their students) has been invaluable. But my greatest debt is to my wife, Marilyn, and family – not only for their putting up with unusual holidays in unconventional locations.

About the author

John Crothers was educated at Solihull School and at St Catharine's College, Cambridge. After graduating BA in 1962, he gained a postgraduate Certificate in Education a year later. In July 1963, he joined the staff of the Field Studies Council as Assistant Warden of Dale Fort Field Centre under the redoubtable John Barrett.

In 1967, John Crothers was appointed Warden of the Leonard Wills Field Centre at Nettlecombe Court in Somerset and remained in post until he retired at the end of 1999.

Academic staff at field centres have endless opportunities for research, and John published more than 75 scientific papers; those relating to the geographical patterns of shell shape variation in dog-whelks earned him a PhD degree from Cambridge University under the Special Regulations in 1985.

In 1977, he succeeded John Barrett as the editor of *Field Studies*, the journal of the Field Studies Council, and he continued in that role until 2003 when the journal ceased publication. Since 1991, he has edited the *Synopses of the British Fauna* for the Linnean Society of London.

He reached the rank of Major (Royal Signals) in the Territorial Army, served ten years as an Appointed Member of the Exmoor National Park Committee and sat on the councils of the British Ecological Society, the Linnean Society and the Malacological Society.

He met his wife, Marilyn, whilst he was at Cambridge and they married in 1964; they have a son and a daughter.

1 Introduction

Rocky sea shores are among the best habitats for natural history investigations. Not only is there public access (once you have got there) but also they are different, exciting and, potentially, slightly dangerous places.

The lives of animals on rocky shores seem to be dominated by physical factors that we, too, may experience – including desiccation, inundation, wave action and extremes of temperature. The effects of these physical factors may change significantly over very short distances so that zonation and other distribution patterns may be instantly apparent. As a bonus, most of the animals and plants live out on the open rock surface so that there is often no need to disturb the habitat in order to observe them. Finally, rocky shores are among the most 'natural' of habitats in the British Isles; unless there has been a recent oil spill and away from outfalls, rocky sea shores are unlikely to have been greatly affected by human activity.

Of the many different kinds of invertebrate animals to be found on British and Irish rocky shores, marine snails (Phylum Mollusca, Class Gastropoda, Sub-Class Prosobranchia) are a particularly easy group to investigate, thanks to the strong hard shell that they secrete to protect the delicate body. Shells are easy to measure, and also to mark in various ways without affecting the behaviour of the snail.

This book is concerned with living snails not their empty shells. (Some people refer to these as 'dead shells' but this is a misnomer; they are the shells of dead snails.) Collecting such shells may be a pleasant, and harmless, pastime but their distribution will not provide much biological information. The composition of shell beaches tells us more about the vagaries of water movements or the resistance of certain shells to the erosive effects of wave action than it does about the differential abundance of the living snail fauna.

Some shells are put to a secondary use after the death of their original owner. Hermit crabs use them as protection for the abdomen, regularly up-sizing their homes as they grow. Adults of the largest hermits almost always end up in shells of the common whelk (*Buccinum undatum*) and the locations of empty whelk shells on the shore may relate more closely to the activities of hermits (or of their avian predators) than to the activities of whelks!

The sea shore is, by definition, the area that is sometimes covered and sometimes uncovered by the sea. Human observations of the invertebrate shore fauna are, not surprisingly, concentrated on the daytime periods of low tide. But the fauna is almost entirely composed of marine species that have colonised the area from the sea-bed beneath the tidemarks. The animals are usually most active at high tide or at night and the day-time low-tide periods are times to be endured – especially in warm sunny weather. Only the most highly evolved species can survive the conditions high on the shore and it is usual to find that species richness increases with distance down the shore.

Any field work involving sea shores is dependent on the tidal cycle operating at the chosen site; and Britain experiences as great a variation in tidal range as is to be seen anywhere in the world. Our largest tidal range, of more than 17 metres, occurs in the Severn Estuary (under the original Severn Bridge) whilst the smallest, of 0.5 metres, is credited to Machrihanish in southwest Scotland (see p. 62). In those parts, waves can be more significant than tides and atmospheric pressure has more influence than the moon on water levels.

Irrespective of the tidal range at your chosen location, the lowest (and the highest) tides, called spring tides, occur shortly after periods of full and new moon and neap tides, those of smallest amplitude, fall at times of the first and last quarters of the lunar cycle. The most dramatic spring tides are seen at the equinoxes and the least impressive ones at the solstices. Over and above the regular annual pattern, there are longer-term cycles that cause very small variations in the highest and lowest water levels – only really noticeable in the Bristol Channel and other areas of large tidal amplitude.

Tidal predictions, originating from the Proudman Laboratory, Liverpool (known irreverently to some as the

Canute Institute), are calculated assuming normal temperature and pressure, but observed water levels are also influenced by variations in atmospheric pressure and in the strength and direction of the wind. High pressure depresses water levels and low pressure allows them to rise higher; strong onshore winds raise levels and strong offshore winds lower them, especially in bays and estuaries.

On most shores, the greatest variety of weird and wonderful sea creatures is to be seen at extreme low water of spring tides. Few marine biologists can resist the call to hunt along the water's edge on such occasions, especially when high pressure and an offshore wind have pushed the tide down further than usual. I expect all readers of this book will be similarly drawn; and why not? But it is unwise to plan any serious investigations on the very low shore. Not only will your time on site be strictly limited but it may be a long time (perhaps years) before the sea goes out that far again.

It is natural for people to be excited by finding an example of a rare species. But, beyond identifying it and making a note of its name and where and when you found it, there are not many questions you can ask of it. The answer to the obvious one, "Why has it turned up here?" is probably "by chance" or "because it made a mistake."

A central theme of this little book will be that, actually, the really common animals are amongst the most interesting. Not only do you not have to spend hours searching for them, when your own ability to find them becomes a major feature of their apparent distribution, but also you can be sure that if they are rare or absent at a particular site there is a very good reason for it.

2 The biology of marine snails

Fig. 1. A limpet shell.

Fig. 2. The shell of a dog-whelk.

siphonal groove

Fig. 3. The shell of a topshell with, inset, a topshell operculum.

[1] **Operculum**: a plate, often ear-shaped or circular, used to close the shell aperture when the animal withdraws inside.

[2] **Umbilicus**: in some shells, the whorls do not touch in the centre of the spiral, leaving an open space; the umbilicus is the lower opening to that space.

On almost all rocky shores around the coasts of Britain and Ireland, the snail fauna is dominated by members of four groups – *Patella* limpets, topshells, winkles and dog-whelks. They are very easily told apart. Limpets have a simple conical shell (fig. 1) and the animal does not have an operculum[1]. Dog-whelk shells have a siphonal groove (fig. 2), so called because the snail extends a siphon or breathing tube along it when moving about. Topshells have a circular operculum and a shell characterised by a nacreous (mother-of-pearl) inner layer (fig. 3). In older individuals, the coloured outer layer is often abraded away, leaving the nacreous layer exposed, especially at the apex. Most species show an open umbilicus[2], at least when young, although in some it closes in later life and its position is revealed by an umbilical scar. Winkles have an oval operculum and a shell with a dark inner layer (fig. 4).

The shell

The shell is formed on a matrix of the protein conchiolin, which is stiffened by the addition of calcium carbonate crystals. In some species, including the common whelk *Buccinum undatum*, the outermost layer of the shell, the periostracum, remains free from calcium carbonate and the hard shell appears to be covered with brown fur.

The first shell formed by the larval snail (see below) is retained as the apex (the highest point of figs 1–4, see also plate 1.1), although it is often abraded away in old individuals. The outer surface of a snail's body, the mantle, is in contact with the inner surface of the shell. Cells at the edge of the mantle secrete the outer layer of new shell along the lip of the aperture (which extends all the way around the rim in limpets, fig. 5). Just below the arrowhead labelling 'mantle edge' are three black marks showing where the snail is accumulating pigment to form darker shell for the decoration. As the snail increases in size it lays down increasingly thick shell. To ensure that the upper whorls also increase in thickness, the snail secretes shell from the whole outer surface of the mantle. So the outer layers are generally thickest near the lip and the inner layers near the apex, which explains why the

Fig. 4. A winkle shell with, inset, a winkle operculum.

outer layer has been abraded from the apex of the topshell in fig. 3.

Food and feeding

Dog-whelks are carnivores, feeding primarily on sedentary barnacles and mussels but occasionally on snails. All the other species of whelk appear to be carnivores or scavengers as well.

Most other snails are herbivores but it is often far from obvious what they are eating. Indeed rocky shores present a conundrum in that the grazing snails may be most abundant on the shores with the fewest visible algae whilst seaweed-covered shores support comparatively few individuals – except of flat winkles. Lacking a vascular system comparable to that of higher plants, algae are not able to move the products of photosynthesis around the plant body as effectively. Essentially, any excess produced by a cell is released into the water as dissolved organic matter. When the alga dies, more dissolved organic matter is released along with some particulate organic matter (detritus). All this material is available for bacteria to consume and so for animals through eating the bacteria, the particulate organic matter, or by direct absorption of dissolved organic matter. The animals never have to meet their food plants!

Flat winkles are unusual amongst these snails in that they feed on the brown fucoid seaweeds, and/or on any epiphytes growing on the surface of the fronds. Blue-rayed limpets, similarly, feed on kelp and *Fucus serratus* but the other limpets, topshells and winkles all graze over the rock surface even though, to us, there appears to be nothing there to eat.

No doubt they would prefer to eat larger food, especially green algae, but under normal circumstances these have all been eaten before they have a chance to grow large enough to become visible to us. From time to time, when conditions are particularly favourable for rapid algal growth, the grazers do not manage to consume all the growth and grazing tracks through the vegetation become visible.

Breathing

In general, prosobranch snails breathe by means of gills. In what are regarded as the most primitive forms

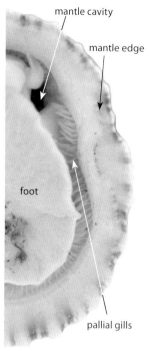

Fig. 5. The left half of a living *Patella* detached from the rock and photographed upside down in sea water.

mantle cavity

mantle edge

foot

pallial gills

alive today, there is a pair of these ctenidia in the mantle cavity. This cavity lies above the head; it is the space into which the head, and often the foot as well, retracts when the snail withdraws under, or into, its shell. In topshells, winkles, and whelks there is but a single gill (ctenidium). *Patella* limpets have lost their ctenidia and, instead, employ a large number of pallial gills, of a different evolutionary origin, which line the groove between the mantle and the foot (fig. 5).

Gills function well under water but are much less useful out of it as their effective surface area is greatly reduced when the lamellae touch each other. One of the problems facing all marine animals that attempt to live on sea shores is how to breathe at low tide. Those poorly adapted to do so are confined to the lower shore (or to rock pools) and only those that have solved the problem are found higher up. Even then, it is probably their ability to breathe and conserve water that dictates the upper limit of their distribution.

Gaseous exchange may also take place through the skin – but only if it can be kept moist. A snail crawling actively over the rocks at low tide on a warm and windy day would rapidly become dehydrated. That is why rocky shore snails are usually most active at night or whilst the tide is in.

Life cycles

Many species of marine snail have separate sexes, but others are hermaphrodite. *Patella* limpets, for example, are protandrous hermaphrodites, maturing first as males before undergoing a sex change and developing into females. It is generally assumed that an adult has to accumulate greater food reserves in order to produce yolky eggs than sperm.

In the forms that we regard as more primitive, including *Patella* limpets and topshells, the males do not have a penis and copulation is impossible. Amongst adults of such species there is no courtship behaviour. Males and females do not even need to meet; they simply release their gametes while they are submerged and fertilisation takes place in the sea water. This appears to be a very wasteful system and it is hard to imagine it being particularly

[3] **Plankton**: organisms floating within a water body. They are usually able to swim but not strongly enough to influence where they are taken by the water body.

Fig. 6. Trochophore larva of *Patella vulgata*.

Fig. 7. Veliger larva of *Littorina littorea*.

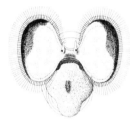

Fig. 8. Part of a breeding aggregation of the dog-whelk, *Nucella lapillus*, with some of their egg capsules.

egg capsules

successful at low population densities. Even where the species is abundant, a degree of co-ordination in the timing of gamete release would appear essential for success.

Living, as they do, close together on the rock, members of a breeding population would all be expected to experience much the same blend of environmental conditions and all would 'come into season' at much the same time. It then needs some special stimulus – for example, the shock of cold water flooding the rock as the flood tide engulfs it after a hot day – to trigger the first individual to release his or her gametes. As soon as another individual detects the presence of gametes in the surrounding water, (s)he releases his or her own gametes and a chain reaction is established.

The free-floating fertilised egg hatches into a planktonic[3] larva; in some species this is a trochophore (fig. 6) that later develops into a veliger. Others hatch directly as a veliger (fig. 7). The time spent as a larva varies greatly between species. Scheltema (1971)* concluded that some veligers can remain planktonic long enough to cross the Atlantic, although not those of any rocky shore species found in Britain or Ireland (the North Atlantic Drift brings us water from the West Indies). At the other end of the scale, in the common topshell, *Osilinus lineatus*, the pelagic stage may last as little as four days. In the edible winkle, *Littorina littorea*, egg and veliger stages may last seven weeks.

Some of the snails that do copulate, including the winkles *Littorina littorea* and *Melarhaphe neritoides*, retain planktonic eggs and larvae. Others, however, lay eggs in masses of jelly attached to fucoid seaweeds (*L. obtusata* and *L. fabalis*), or rocks (*L. arcana, L. compressa*) or in capsules attached to the rock (various whelks including *Nucella lapillus*, fig. 8 and plate 1.6). In all these cases, the larval stages are completed within the jelly mass or capsule and junior emerges as a tiny snail, known as a crawlaway, into the habitat chosen for him (or her) by the mother. *Littorina saxatilis* females take this a stage further; the mother retains the eggs within her mantle cavity and, apparently, gives birth to live young.

---

* References cited thus in the text appear in full in the Further Reading list on p. 83.

Many features of the biology of rocky shore snails can be related to the form of their life cycle. The wastage incurred by species with planktonic larvae must be enormous. Not only are the larvae vulnerable to predators, but large numbers will find themselves dumped in unsuitable habitats. Yet those that survive may have been distributed over a wide area. It is easy for such species to recolonise sites following an event such as an oil spill, a cold winter or a hot summer, that reduced the population. They may also be well placed to extend their range northwards if temperatures rise.

Those more protective snails, whose young begin their lives in the habitat chosen for them by their mothers, might be expected to suffer from much lower infant mortality but to have more difficulty in extending their range. However, recent events have shown them to be less restricted in this way than was once imagined. Shores depopulated by various 'disasters' have been naturally repopulated.

3 Limpets

Fig. 9. At low tide by day, a limpet rests 'at home' surrounded by evidence of its past feeding forays.

Fig. 10. This limpet had crawled forward, swinging its head from side to side, scraping the (pale) algal covering off the (darker) rock surface.

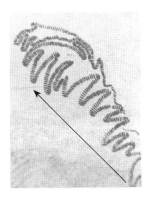

Fig. 11. In close-up, the track is seen to be formed of discrete radula scratches.

The common limpet, *Patella vulgata* (plate 1.1), is one of the most characteristic and ubiquitous animals on British and Irish rocky shores; it is also one of the most important because it controls the growth of seaweeds. But it has failed to make much of an impression on human visitors to the shore, being active when it is dark and wet whereas most people see sea shores in daylight when it is dry and sunny. Very few limpets 'do anything' at low tide by day and so they never attract the attention given to crabs or fish. If anything, their value as subjects for investigation is enhanced by the lack of interest shown by people in general because human interference is unlikely to be a complicating factor in most cases. *Patella* is almost impossible to confuse with any other genus of snails and individuals are easy to find, count, measure and mark. They also leave interesting tracks on the rock.

Grazing tracks

At low tide by day, most limpets are inactive, but they are often surrounded by tracks which record their nighttime grazing activity. In the photographs (figs 9–12), the tracks show up where the snails have scraped a pale-coloured algal film from off the dark grey rock. Close up (figs 10 and 11), each track is seen to be made up of discrete, parallel-sided marks (radula scratches or tongue licks) formed as the limpet crawled slowly forward, swinging its head from side to side, rasping the algal felt off the rock. In fig. 10, the individual has grazed from bottom right to top left of the frame and then has swung round to its right and set off back again by a more direct route.

It is likely that a limpet makes only one trail a night. So the animal illustrated in fig. 9 appears to have a 'home' on the rock to which it had returned before each new day. The most recent tracks are the most clearly defined; older tracks become blurred as the algae grow back, in much the same way that mower stripes fade away on lawns.

Note that this limpet has taken a different route every night. By so doing, it is managing its food

Fig. 12. Limpet grazing tracks radiating out from the individual's 'home' on the rock. Recent tracks are clearly defined, older ones less so. The track indicated by white arrows was made on a different occasion to that indicated with black arrows. The gaps in some tracks reveal that limpets were not feeding continuously when crawling. (An enlargement of part of fig. 9.)

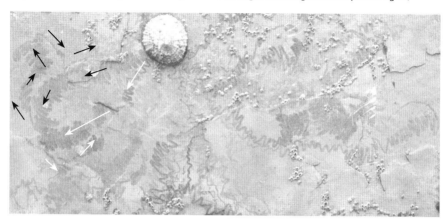

Fig 13. Limpet homes clustered together on Watchet harbour wall.

Fig 14. A more distant view of limpet-grazed areas on the wall, each with a cluster of homes.

resource. If it always set off in the same direction, it would soon scrape the rock clean along its path and have to crawl ever further each night to find a meal.

Many people have wondered how a limpet knows when to stop feeding and turn for home, and how it finds its way back. All snails secrete a mucus trail to crawl over. I suspect that the trail left by a limpet is thinner that those left by the much more active topshells and winkles because I have never seen one. That does not mean that the limpet cannot recognise its own trail and follow it home. But it does not slavishly retrace its outward track. In fig. 12, the black arrows indicate an occasion when the limpet, intelligently, took a short cut on the way home!

Tracks are only visible here because most of the algae had not been harvested; where limpets are very numerous (or when algal growth is slow) no tracks may be visible. Fig. 13 illustrates limpets living on the steep concrete wall of Watchet harbour, in Somerset. A year earlier, the wall had been re-surfaced; as all the resident grazing snails were removed by this process, it was rapidly colonised by green algae and then, in some places, brown algae settled on the green algal felt. Limpets re-colonised the wall from the rock below and the pale area in the photograph shows where the concrete has been grazed clean of algae. As time passed, the cleared areas increased in size (fig. 14) until they coalesced; eventually, the whole wall was cleared of visible algal growth.

Fig. 15 shows a small rock pool on an exposed rocky shore. The pool bed appears paler than the surrounding rock surface because it has been colonised by a coralline pink alga, *Lithothamnion*. (Coralline algae are not grazed by limpets or other snails.) Some of the largest limpets show up prominently because they have tufts of green algae growing on their shells. This is a very common phenomenon, which has prompted much speculation regarding its cause. Does the limpet benefit from the oxygen released into the water from the photosynthesising algae? Do the algae offer camouflage? They make the limpets more conspicuous to us but might an oystercatcher be confused? However, amidst all the speculation, it is important to remember that an algal spore is not able to choose where it will settle; they are broadcast everywhere.

The question is not "Why are the algae growing on the limpet shells?" but "What has happened to the algae that settled everywhere else?" The answer is that they have been eaten – probably by other limpets! Presumably a limpet does not welcome another snail crawling onto its shell; it might be a predator (see p. 27). However, this does sometimes happen and fig. 16 illustrates a limpet shell with grazing marks running across it.

Fig. 15. Tufts of green algae growing on the shells of limpets in a rock pool.

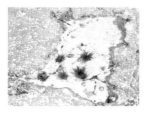

Fig. 16. A limpet, at home, with scratches from the radula of another snail on its shell.

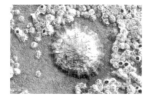

Homing behaviour

Homing behaviour is a characteristic feature of limpets in the genus *Patella*. It is generally observed that a limpet's shell fits the rock surface of its home very closely – far too closely for this to have occurred by chance. On soft surfaces, the snails grind their shells into the rock to make it fit their shape, forming 'scars' (fig. 17). The number of empty limpet scars might indicate the scale of recent limpet mortality, after an extreme weather event or pollution incident, but there are always a number of empty limpet scars on every shore where the rock is soft enough for them to be formed – and new recruits to the limpet population often colonise them preferentially.

Fig. 17. Two limpets at home with a scar that had been ground into the soft rock by a third.

limpet scar

Fig. 18. A limpet with a home on hard rock must grow the shell margin to fit the rock.

former edge of the shell

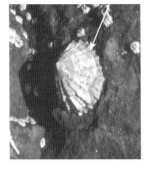

Fig. 19. A living *Patella* detached from the rock and photographed upside down in sea water.

mantle   foot

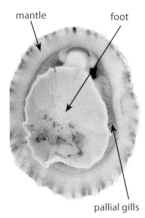

pallial gills

On hard igneous or metamorphic rocks, a limpet could spend a lifetime grinding without affecting the rock very much. On such surfaces limpets grow their shells to conform to the rock. In this example (fig. 18), a considerable amount of new shell material has been laid down in order to achieve a close fit with this particular piece of rock surface.

Why is it important that the shell fits the rock surface so closely? The answer is, probably, in order to conserve water, enabling limpets to breathe using gills on the open, dry, rock surface close to their feeding site.

As we have seen, *Patella* limpets breathe by means of pallial gills lying in the groove between the mantle and the foot (fig. 19). Thanks to the close fit between shell and rock when 'at home', water is held in this groove, by capillarity. When the rock dries out, and after very slightly relaxing its hold on the rock, the limpet can establish an air-water interface all round the margin of the shell through which oxygen may diffuse in and carbon dioxide diffuse out during the low tide period (fig. 20).

Before the rock dries out, limpets return to their homes, clamp down tightly until the risk of dislodgment by wave action is past and then relax to replenish the oxygen supply in the water under their shells. It is then that they are vulnerable to a sharp sideways tap by, say, an oystercatcher, *Haematopus ostralegus*.

Major disruption of limpet populations is usually followed by recolonisation of 'damaged' sites by adult limpets (for example, see Crump, Williams & Crothers, 2003) and the individual illustrated in fig. 18 had colonised its new home following an oil pollution experiment that had eliminated the existing limpets; it rapidly produced more shell to conform to the rock surface. But in an undisturbed location, homing had been generally assumed, for this is such a beautiful example of the co-evolution of behaviour and morphology.

However, this is not always the case on all shores. Pupils at Selwood Middle School, from Frome, Somerset, carried out a 5-day investigation, one June, that involved marking limpets (as in fig. 21), using

Fig. 20. Limpets breathing whilst out of water at the edge of a rock pool.

Fig. 21. A limpet marked for an experiment designed to monitor homing behaviour.

Fig. 22. Limpets clustering on Watchet harbour wall.

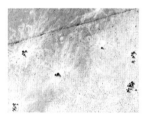

Fig. 23. A limpet that must orientate itself very carefully to fit into its home.

different colours of paint on successive days. They found that only 60% of the animals were in the same home every morning. Some had two homes, others more.

The middle shore at Watchet, Somerset, where they carried out their investigation, is formed from Jurassic Tea-Green Marl that has eroded to form an extensive wave-cut platform. It is not exposed to much wave action, and in most years it supports a good cover of fucoid brown seaweeds (see p. 64).

It could be that homing behaviour is more important on the upper shore than on the middle or lower shores. It may be more important on rough surfaces than on smooth, on steep slopes rather than on nearly horizontal surfaces and on exposed shores rather than sheltered ones.

Where the upper shore is a cliff, or a wall, the limpet homes are often clustered (fig. 22) much as woodlice cluster in dry conditions. It has been hypothesised that this behaviour may help the limpets to retain moisture at low tide by day. Experiments, using empty limpet shells filled with wet blotting paper, have been used to test this suggestion, with encouraging results. It is currently unclear whether the clustering habit is associated with height up the shore (equivalent to length of time out of sea water) or with the steepness of the rock. Fig. 22 shows the same harbour wall as shown in figs 13 and 14, a few years later when most of the visible algae had been grazed off.

Orientation of limpets in their homes is another variable worthy of investigation; sometimes local topography seems to be the most important factor (fig. 23) but on other occasions almost all individuals will be found facing in the same direction. Are most of them head downward in their homes? (The head is under the apex of a *Patella* shell; that is to the left in the central individual of fig. 20). Is the pattern related to the steepness of the rock? or height up the shore? Do limpets usually face into the waves?

Biometrics

Baby limpets arrive on the shore less than 0.5 mm long and have the potential of reaching 60 mm in sixteen or more years.

There is usually an inverse relationship between limpet size and abundance; a fixed amount of food will support more small individuals than large ones. Generally speaking, exposed, barnacle-dominated shores support high densities of small limpets whilst sheltered seaweed-covered shores harbour lower densities of larger ones.

The big brown seaweeds exclude most of the light from the rock surface, shading out the diatoms and other components of the algal felt upon which limpets graze. Moreover, by the whiplash effect of their fronds on the rock (fig. 24) they abrade any remaining small algae (and limpet spat[4]) from their area of influence. The limpets found under seaweeds either were living there before the seaweed arrived or had settled somewhere else and migrated under the canopy once they were large enough to withstand the whiplash. Both Boaventura, da Fonseca & Hawkins (2003) and Crump, Williams & Crothers (2003) demonstrated that it was shortage of food and not wave action that kept exposed shore limpets small. If density is reduced, or the food supply improves, the survivors grow rapidly.

Often it is observed that the mean size of the limpets varies as you move up or down the shore – but the maximum size hardly changes. Variation in the mean depends largely on the number of small individuals present. It seems likely that there are nursery levels for, whereas limpet spat are broadcast across the entire shore, survival of baby snails is much more restricted. Shortage of food due to overcrowding, after a successful spatfall, will encourage the youngsters to move off as soon as they can tolerate conditions away from the nursery.

Upper-shore limpets often have more highly conical shells (fig. 25) than those on the middle shore. Traditionally (Orton, 1929), this was explained as the effect on growth of the inward pull of the shell muscle to keep the shell close to the substratum for the long

Fig. 24. To illustrate the whip-lash effect of a clump of fucoid alga.

area of the rock surface swept by this clump of saw wrack

[4] **spat**: a general term for larvae newly settled from the plankton and the very young molluscs that have just metamorphosed from them.

[5] See p. 77 for techniques for measuring limpets without harming them.

Fig. 25. Upper-shore limpets on granite, Isles of Scilly.

Fig. 26. Shells of limpets taken above high water mark by gulls.

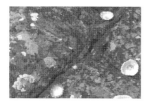

Fig. 27. Shell of a limpet eaten on the shore, probably by an oystercatcher.

Fig. 28. Large numbers of empty limpet shells in the strand line litter after a few very cold nights in winter.

Fig. 29. A desperate attempt at cooling on a hot day?

periods of low tide. However, limpets do not spend all the low tide period clamped hard down, but relax in order to breathe. Boaventura, da Fonseca & Hawkins (2003) found that limpets living at low densities, with a good food supply, grew rapidly and developed rather flat shells. Does that suggest that the high conical shells indicate slow growth, perhaps due to shortage of food or of time available for feeding?

Bird predation

Gulls, *Larus* species, and oystercatchers, *Haematopus ostralegus*, are able to knock limpets off their homes when they are relaxed, breathing. The bird then often wedges the shell in a crevice before pecking out the meat. Gulls often fly up above high water mark to do this (fig. 26), but oystercatchers usually feed near where the prey was found. If clean, inverted, limpet shells are found on the top of rocks, birds have been feeding since the tide went out (fig. 27).

If the birds were taking prey items at random, the size-frequency distribution of predated limpet shells should be the same as that of the survivors. This is rarely the case because the birds select within a preferred size range. Do they favour the largest available prey, the modal (= commonest) size, or some other subset of the population?

Unusual mortality

Occasionally, when visiting a shore, one realises that some natural disaster has recently occurred that has caused the death of an unusually large number of limpets. Fig. 28 shows the sequel to a period of unusually cold nights. Heatwaves that coincide with periods when low water falls around noon may also have significant effects.

Fig. 29 is a photograph of a limpet suffering from severe heat stress. It was living on a south-facing rock surface in full sun on a day so hot that it was uncomfortable to stand on the rock with bare feet. The animal has rolled up the edge of its foot, presumably to reduce contact with the rock surface, and raised the shell clear of the rock. This would have aided cooling by evaporation whilst increasing the risk of desiccation. Not surprisingly, it was the limpets exposed to full sun that were most obviously distressed.

Where an observed high mortality of limpets can be related to a recent incident, natural or man-made, it would seem to be a simple matter to measure the size frequency distribution of the casualties and compare it with that of the survivors. When interpreting the data, however, consideration should be given to the influence of longshore drift and also to the differential resistance of limpet shells to erosion; large shells persist for longer than small ones.

Listening to limpets

I wrote above that the rock-dwelling *Patella* limpets do not crawl about very much when the rock is dry. I should not have suggested, however, that all remain totally inactive at low tide by day.

On those shores where it is possible to walk into caves or between the walls of ravines, the damp, shaded lower shore can be a noisy place. Some of the noise is due to barnacles and bivalves blowing bubbles but most is due to limpets. Whether they are feeding or grinding their shells against the rock, I do not know.

Blue-rayed limpets

Patella pellucida (= *Patina* or *Helcion*) is a very different beast, usually found grazing on red seaweeds, saw wrack (*Fucus serratus*), or kelp, especially *Laminaria digitata*. The largest individuals are usually found within the kelp holdfasts, living in cavities that they have scraped out of the alga.

They live on the lower shore and in the shallow subtidal, so they are only accessible at spring tides and different areas of their habitat are exposed on successive days. This makes it difficult to investigate their biology and behaviour, but the handsome iridescent blue flashes on the shell make them well worth searching for.

The best sources of information on basic limpet biology are Fretter & Graham (1962, 1976, 1994). Limpets are excellent subjects for seashore investigation, but there are two major drawbacks; they are difficult to identify to species, and to age.

4 Common topshells

Fig. 30. *Osilinus lineatus*

Population studies

A more tempting subject for investigation is the common topshell, *Osilinus lineatus* (fig. 30, plate 1), which is easy to identify (in Britain and Ireland), and which lives on the upper middle shore and so is available on all tides. Further, it is possible to age most members of most populations.

It is at the northern limit of its geographical range in southwest Britain and western Ireland (see Crothers, 2001). Presumably that limit is reached when winters are too cold, or too cold for too long, and/or summers are not warm enough, or not warm enough for long enough. The cold winters of 1947 and 1963 pushed the limits south (and west, in the English and Bristol Channels); the higher water temperatures experienced during the last three decades have allowed the limit to extend north (and east). On our shores, growth is usually suspended for several months each winter. During this time, the edge of the shell becomes worn, chipped or damaged so that, when growth begins again the following spring, a 'conspicuous' growth check (fig. 31) is seen on the outer surface of the shell. Counts of these growth checks give an estimate of the number of winters the snail has survived (Williamson & Kendall, 1981).

As breeding is largely confined to a short period in mid-summer, the population is composed of discrete cohorts. Young snails appear on the shore in August or September each year. Juveniles grow rapidly to reach adulthood; adults continue to grow much more slowly for the rest of their lives, which may exceed 15 years.

Fig. 31. *O. lineatus* showing the 'conspicuous' growth checks that represent the ghosts of winters past.

The annual growth checks cross the outer surface of the shell parallel to the outer lip but are most clearly defined where they cross the white area underneath (arrowed in fig. 30). In old individuals (plate 1.5), the recent checks may be quite close together in contrast to others now higher up the spire which were formed when the young snail was growing rapidly (plate 1.4). In such individuals the outer, coloured layer of shell is often abraded from the apex, removing all trace of any growth

conspicuous
growth checks

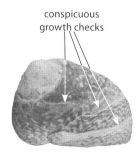

Fig. 32. Old *O. lineatus* at the edge of a rock pool.

Fig. 33. An *O. lineatus* that had survived a crab attack and repaired the damage by laying down more shell.

Fig. 34. I took the shell length of *O. lineatus* to be the maximum distance between the apex (left) and the opposite margin of the shell lip.

check relating to the animal's first winter – so its presence should be assumed, even if it cannot be seen.

However, 'conspicuous growth check' is a technical term and they are not always that obvious – especially after a mild winter when growth was not totally suppressed. In such circumstances, many shells show a band of growth checks that are better described as 'barely-discernable' than 'conspicuous'. The most difficult season in which to age these snails is from May to July, because young snails begin to grow much sooner than older ones and so appear to have an additional check.

Not all growth checks are due to low winter temperatures; damage to the shell can also check growth (fig. 32 right). Damage in winter is probably caused by storms, but chips out of the lip resulting from crab attack (fig. 33) probably occurred in summer when large crabs are more active on the shore. It might be possible to estimate the comparative importance of (unsuccessful) crab attacks on different shores by comparing the proportions of living snails which bear such injuries. Successful attacks could be estimated from the number of empty shells with large chips missing from the shell lip. For information about shore crabs, see Crothers (1967, 1968).

Growth rates may be established by calculating the mean size of each year class. It is probably most useful to measure shell length – the distance between the apex and the opposite edge of the shell lip (fig. 34) – but some people have used basal diameter; they are numerically similar figures in this species.

On Gore Point growth rates appeared remarkably consistent between 1994 and 2000 (fig. 35C). Each new year-class settled on the shore in late August or early September at a mean shell length of around 2 mm. By the end of October they had reached some 5 mm and growth ceased for the winter. Then followed the maximum period of growth as the snails passed a mean of 10.5 mm by their first birthday to reach 12 mm in November. The cohort's second winter cessation of growth lasted until June when a third burst of growth took the mean length up to 16 mm for their second birthday. Growth then slowed right down so that on their third birthday each cohort averaged around

Fig. 35. Growth pattern of *O. lineatus* on Gore Point, Somerset, from 1994 to 2000.
A shows the passage of one cohort, plotting the mean shell length against the months.
B adds two more and C shows the whole pattern. The site was visited by groups of students
at approximately weekly intervals between spring and autumn and by myself, monthly,
in winter. Samples, collected without conscious bias, were sorted into age classes; each
age class was measured and a mean shell length calculated. When, as occasionally
happened, no data were collected in one month, the figure plotted is the average of the
adjacent months. (The gap marks a time when I was recuperating from a hip operation.)
From Crothers (2001).

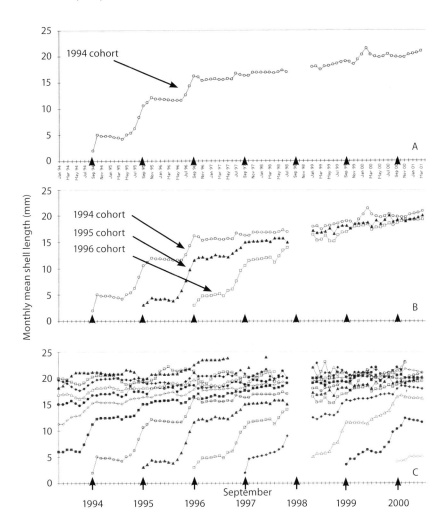

Fig. 36. Grazing track of *O. lineatus* (central snail).

Fig. 37. Unlike limpet tracks (fig. 12) topshell tracks do not radiate out from a central point.

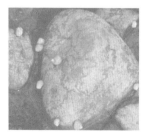

Fig. 38. Grazing tracks of *O. lineatus* and *L. littorea* over a boulder. The broader tracks would seem to be those of the topshell.

16.5 mm and it was not until their seventh that the mean length had passed 20 mm. The most likely explanation for the marked decrease in the rate of growth is the onset of sexual maturity after which a considerable quantity of energy and nutrients are committed to reproduction. Note that, in this population, individuals do not reach 25 mm.

Very young juveniles are difficult to find, so mean values were derived from comparatively few individuals; as was also the case for the cohorts more than five years old. It will have been noticed that, in such cohorts, average sizes show dips as well as rises. It is possible for shells to get smaller, through abrasion, but it is more likely that the aberrant points were due to chance; after all, these were sample means not measurements of the whole population. Perhaps selective loss of larger individuals caused a decline in the mean; individuals who had grown faster than their contemporaries may have been less fitted to survive the following winter.

O. lineatus was present on Gore Point in the early 1940s (Bassindale, 1943) but was not seen anywhere else in Somerset (Bassindale, 1941). The effects of the 1947 cold winter do not appear to have been recorded but those of February–March 1963 eliminated the species from the Bristol Channel coast of Devon and Somerset. By 1967, the species had returned to Gore Point but had not reached much further east. As sea temperatures rose during the following years, *O. lineatus* extended its range eastward along the Somerset coast, reaching Hinkley Point at the head of Bridgwater Bay by 2003.

The age structure of the newly-founded populations was notably different from that of the established population on Gore Point. In the new enclaves, the snails had reached 7 mm by the first autumn and 17 mm by their first birthday; animals of 20 mm were apparently in their third year. Individuals of over 25 mm length were not unusual.

As with limpets, it seems that where *O. lineatus* is comparatively rare, individuals grow rapidly to a large size, appearing to reach maturity a year earlier, but may not survive to any great age; where the species is really common, individuals grow slowly, do not achieve any

Fig. 39. *O. lineatus* is capable of clearing areas of algal felt off the rocks. Compare those cleared by limpets, fig. 14.

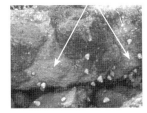

grazed areas

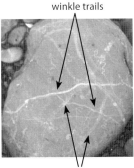

Figs 40 and 41. Slime trails. Winkles leave a continuous band of mucus, topshells a more distinctive trail.

winkle trails

topshell trails

great size but may live to a ripe old age. It seems likely that size and rate of growth are limited by the food supply whilst abundance is controlled by the successful settlement of spat and subsequent evasion of predators. As nobody has discovered any major difference in the food preferences of *Patella* species, *Osilinus lineatus*, *Gibbula umbilicalis*, *Littorina littorea* and *L. saxatilis*, it is probably the combined density of these rock-scraping snails that matters.

Behaviour

Common topshells and edible winkles (*Littorina littorea*), like limpets, may leave grazing tracks on the rock surface (figs 36–38). Like *Patella* (figs 10 and 11), a grazing *Osilinus* swings its head from side to side while crawling, but the whole process is much more rapid in this species and there is not time to swing the head so far.

Like limpets, common topshells can clear an algal felt off the rock (fig. 39), but it is usually less obvious that they were the cause.

In addition to leaving the radula scratches of grazing tracks, topshells and winkles also leave visible slime trails. Actually, all snails leave mucus trails, but those of *Patella* are not conspicuous to humans. The more active topshells and winkles presumably secrete a thicker layer on which to crawl and this may show up as a pale band over the rock surface. Snails crawl by squeezing the front end of the foot against the substrate and, by means of a ripple of muscle contraction, pass that point of contact aft across the sole thereby forcing the mass of the snail forwards. If you can persuade a snail to crawl across a glass plate, watch the process through the glass. You will usually see several waves of contraction moving at once.

In littorinids (winkles) the foot musculature appears to work as a whole and the mucus trail appears as a plain pale band on the rock (fig. 40). In topshells, however, the two halves of the foot work independently of each other, out of phase, producing a characteristic slime trail (fig. 41).

Limpets may be inactive at low tide by day, but the same is not true for common topshells. This can be a serious concern during monitoring surveys because

Fig. 42. Adult *O. lineatus* on the top of a boulder on a bright sunny day. (The pale, alga-free, zone around the barnacles will be discussed under the rough winkle *Littorina saxatilis*. p. 41)

Fig. 43. The freshwater stream on Gore Point at low tide. Common topshells and edible winkles have crawled out of the low-salinity water onto the top of cobbles and boulders.

counts of this species vary with the weather. Little, Dicks & Crothers (1986) observed topshells to respond within minutes to changes in weather conditions, moving out into the open when the sun shone (fig. 42) but hiding from rain or cold winds in gullies or under boulders.

As with limpets, there appear to be nursery areas in that small juveniles are much commoner in certain parts of the shore than in others, but there appears to be no uniform pattern; sometimes the site is low down, sometimes higher up. Perhaps it is related to particle size of the substrate, perhaps to the availability of pools of an appropriate size. On some shores, large individuals migrate upshore in summer and retreat downshore in winter.

On Gore Point in Somerset, a small freshwater stream runs across the beach. Although *Osilinus lineatus* is no more tolerant of fresh water than are the other species of marine snail, it is more abundant than the others in the area of the stream (p. 72). Further, it reaches its highest density in that area. Animals marked on another part of Gore Point (p. 77, 79 or 80) moved into the stream area; snails in the stream area grow larger and appear to live longer, although no first year individuals (0+ age-class) have been seen there.

None of the grazing snails (limpets, topshells and winkles) can feed in fresh water, so during the low tide periods the algae living there are protected and their herbivores must move out onto the dry areas which are quickly grazed bare. When the flood tide returns the snails wake up and return to feed. As the ebb sets in and the fresh water begins to wash out the sea water, adult *O. lineatus* (and *Littorina littorea*) have time to crawl clear of the stream (fig. 43) to avoid the osmotic effects of low salinity. Limpets and small individuals of other species must stay away from the stream or perish.

5 Dog-whelks

Fig. 44. *Nucella lapillus.*

Fig. 45. *N. lapillus* on barnacles.

Fig. 46. *N. lapillus* on mussels.

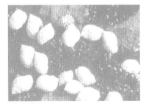

Fig. 47. Holes bored by juvenile *N. lapillus* in the plates of barnacles.

The common dog-whelk, *Nucella lapillus* (fig. 44), is another excellent subject for study. These snails are conspicuous, of a convenient size (with shells usually between 20 and 35 mm long), comparatively long-lived, harmless to man, of no commercial importance, widely distributed and common (except along parts of the south coast of England, p. 37). The shell is usually robust and individuals are easily marked – you can write on the shell with a pencil, paint numbers on it or saw grooves into the aperture. The equivalent of larval stages are completed within an egg capsule so that the young emerge at the crawling stage when they are visible to the naked eye. They grow for two or three years but do not increase the size of the shell as adults.

At all ages, dog-whelks feed on discrete, easily identifiable, macroscopic sedentary or slow-moving prey (upon which they remain for many hours or days) which renders predator/prey investigations possible in the field. Above all, they lend themselves to a study of variation and the morphological response of a species to environmental selection processes.

Feeding

On all but the smallest prey, *Nucella* must penetrate the victim's shell before paralysing it by injecting a narcotic. The dog-whelk then inserts its proboscis into the body of the prey, secretes digestive enzymes and later resorbs a nutritious 'soup'. This process of external digestion means that *Nucella* is a liquid feeder and produces very little in the way of solid faeces. The tiny faecal pellets give no useful indication of the food ingested.

On most shores, *Nucella lapillus* is a predator of barnacles (fig. 45) and/or of mussels (fig. 46). A large dog-whelk can inject the narcotic between the opercular plates of a small barnacle to relax the opercular muscles, and so does not have to bore a hole in the shell. But small dog-whelks feeding on large barnacles do bore holes (fig. 47). They often attack through the opercular plates but sometimes bore through the, much thicker, wall plates near their base. Perhaps the proboscis is not

Fig. 48. A hole bored through the shell of *Mytilus edulis* by *N. lapillus*.

borehole

Fig. 49. A dog-whelk could bore a hole anywhere on the surface of a mussel. But its proboscis can only be extended so far (about the length of its shell). After Carefoot (1977 Fig. 127).

long enough to reach the most succulent parts of the prey by any other means. All dog-whelks have to bore all but the smallest mussels (fig. 48).

The extended proboscis is said to be roughly the length of its owner's shell (fig. 49). If a good supply of bored shells is available, it could be interesting to investigate where, on the prey shell surface, the predator chose to bore the hole. A mussel shell is thinnest at the margins and thickest at the umbo (the oldest part). It is easiest to bore through near the margin but the main organs of the body may then be beyond the reach of the proboscis.

Of course, when the mussels are growing in a dense 'bed', the only area of the shell accessible to the dog-whelk may be that close to the margin. Towards the southern end of its geographical range, in Portugal, *N. lapillus* lives under the mussel clumps rather than on top of them. Initially I interpreted this as a behaviour to avoid the heat of the sun at midday, but I now wonder whether it has more to do with access to the energy-rich organs of their prey.

At one time, it was thought that boring was achieved simply by mechanical scraping with the radula, until Carriker (1981) described how the process involved a combination of chemical and mechanical activity. Chemical softening is effected by the secretion of enzymes (particularly carbonic anhydrase which attacks conchiolin, the protein that forms the organic matrix of the shell), from an accessory boring organ situated in the sole of the foot towards its anterior end. The smooth round shape of the resulting hole (fig. 48) reflects the shape of the accessory boring organ and not that of the radula, which plays a comparatively minor role in the mechanical removal of shell fragments although it may be used to determine when the hole is complete.

Accounts of the boring behaviour (for example Fretter & Graham, 1985) describe how the animal alternately places the accessory boring organ in the hole and then makes a few rasps with the radula. Fretter & Graham (1962) quote figures that suggest a rate of boring into limpet (*Patella*) shells of 0.175 mm per hour but no details are given of the relative sizes of predator and prey, or of the temperature.

Prey selection

Many books claim that *Nucella lapillus* favours barnacles over mussels as food. This often appears to be the case, especially in northern Britain, but further south the situation is more complicated.

The null hypothesis (p. 66) must be that dog-whelks do not select their prey but simply eat each prey item that they come across. Very few animals actually do this. Theoretically, natural selection should have favoured those that fed most efficiently; that is they chose the prey that gave them the most energy per unit effort expended whilst feeding. Carefoot (1977) found that, in aquaria, Pacific *Nucella lamellosa* gained an average of 23 calories (96 joules) per hour from feeding on barnacles (*Balanus glandula*) but only 15 calories (63 joules) per hour from feeding on mussels (*Mytilus edulis*).

We might hypothesise that the dog-whelks would exert apostatic selection, feeding disproportionately on the most abundant prey species available. Having enjoyed a meal on an individual of species *a*, the predator should look for another individual of *a* rather than risk disappointment from attacking an individual of species *b*. In the absence of any individuals of species *a*, and feeling hungry, the snail would be likely to switch to an individual of the most abundant species still available. Over time, all members of the population should end up searching for what was, recently, the most abundant prey species.

In addition to mussels, *Mytilus edulis* (fig. 48), dog-whelks may have up to four middle-shore species of barnacles available as potential prey, plus more on the lower shore. *Semibalanus balanoides* (fig. 50) is probably the most widely distributed of the mid-shore barnacle species and may be the only species present in the north and east of Britain. It becomes increasingly rare in the far southwest where it may be absent from south-facing shores. Species of *Chthamalus* are found on the more exposed shores in the south and west of Britain, with *C. stellatus* (fig. 51) favouring the most oceanic sites and *C. montagui* (fig. 52) slightly less exposed situations. Introduced to the Solent from Australasia during the Second World War, *Elminius modestus* (fig. 53) continues to spread, and seems to thrive best on sheltered shores and in estuaries.

Fig. 50. *Semibalanus balanoides.*

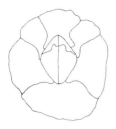

Fig. 51. *Chthamalus stellatus.*

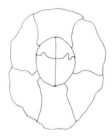

Fig. 52. *Chthamalus montagui.*

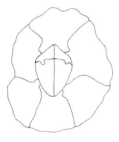

Fig. 53. *Elminius modestus.*

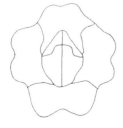

Other, lower-shore and sublittoral members of the barnacle family Balanidae show a similar arrangement of shell plates to fig. 50. Southward (2007) gives further information about barnacles, and identification keys.

The four species illustrated here can be distinguished by examining the wall plates, if visible, and the shape of the opercular opening. *Elminius modestus* (fig. 53) has four plates in the wall and a diamond-shaped opening. The other balanids (including *Semibalanus*) also have a diamond-shaped opening but have six, unequally-sized wall plates (fig. 50) with two small ones at the anterior end. Plate boundaries are often clearly visible, unlike in *Chthamalus* species where the plates fuse together in adults. *C. stellatus* has an oval aperture (fig. 51) and *C. montagui* a kite-shaped one (fig. 52).

My impression is that *Semibalanus balanoides* is the favoured prey item, so much so that in Shetland and Norway (where this is the only midshore barnacle species present) it is not unusual to find that the intense predation has restricted adult barnacles to a narrow zone high on the rocks, where they are safe from attack (fig. 54). When they have consumed all of their favourite prey within their feeding zone, the dog-whelks usually switch to feeding on mussels.

In southwest Ireland, *N. lapillus* preys on mussels in preference to *Chthamalus* barnacles (fig. 55).

Mention has already been made of a site on Gore Point, Somerset, where a small stream flows across the shore. Here, mussels are confined to the area influenced by the fresh water but field observations at low tide always show the shell valves to be clamped shut; they are not feeding in the fresh water. Laboratory experiments have confirmed their inactivity in water of reduced salinity. The explanation for their observed distribution is that *N. lapillus*, too, is unable to feed under these conditions and retracts into its shell. As it takes the whelk much longer than one period of high tide to bore through a mussel shell, the stream offers the mussels a refuge from dog-whelk predation. Absence of mussels away from the stream suggests that *N. lapillus* prefers to feed on mussels rather than on *Elminius modestus* – the commonest barnacle on this shore.

Fig. 54. In Shetland, intense *N. lapillus* predation confines *S. balanoides* to the top of the shore.

Fig. 55. In SW Ireland, *N. lapillus* prefers mussels (dark shells) to *Chthamalus* barnacles.

Fig. 56. In the absence of barnacles and mussels, *N. lapillus* will attack limpets. . .

Fig. 57. . . . and even hunt through the seaweed for flat winkles!

Littorina obtusata

So, maybe dog-whelks favour *Semibalanus balanoides* over *Mytilus edulis* over *Elminius modestus* over *Chthamalus* species; other balanid species, if available, fall somewhere in the middle.

Around the Isles of Scilly, and maybe elsewhere, one may find shores with a dog-whelk population but no barnacles or mussels. At such sites, *N. lapillus* attacks limpets (fig. 56), flat winkles (fig. 57) and purple topshells. Cannibalism has been observed. I have yet to see them attack an edible winkle, *Littorina littorea*, in Britain (although I have in Canada, fig. 58) and have never seen a common topshell attacked. However, whilst this book was in preparation, Dr Barry Shurlock showed me a bored shell from the Gulf of Morbihan; the dog whelk had bored into the body whorl from from inside the aperture!

Snails are often attacked through the operculum (fig. 57), so the fact that empty shells lack boreholes is not positive evidence that they had avoided dog-whelk predation.

Whilst it is comparatively easy to see if a dog-whelk is feeding on a limpet or a mussel (figs 55 and 56) it is not always easy to tell which species of barnacle is under attack.

Under optimal conditions, the predators may consume an average of one barnacle per day so direct observations or experiments will occupy several days; consumption of a mussel may take a week or more. Live coverage of *Nucella* feeding would not make riveting television! Any serious investigation into prey selection and food consumption in the field would involve repeat visits to the same site over a period of days or weeks.

If sufficient time was available, it would be straightforward to set up choice experiments, in the field, for groups or for individual *N. lapillus*. One could merely observe which species were killed and eaten (and which were not) or, by selectively sacrificing some of the prey items, weight the choice experimentally.

It should be remembered that the predator may select prey items on the basis of size, as well as of species, and that the preferred size of prey will probably be related to the size of the predator. Obviously, larger

Fig. 58. Edible winkle shells bored by *N. lapillus*. Cape Enrage, Bay of Fundy.

Fig. 59. An adult *N. lapillus* feeding on newly-settled *S. balanoides*, in Shetland, in preference to barnacles one or more years old. Some opportunistic barnacle spat have settled on the dog-whelk's shell!

Fig. 60. Part of a breeding aggregation of (the unusually elongated) Bristol Channel dog-whelks with recently-laid egg capsules. This photo is reproduced in colour as plate 1.6

prey items contain a greater quantity of food but the shell will also be thicker so it will take longer for the dog-whelk to obtain it. My impression is that dog-whelks choose small prey items when these are available. In fig. 59, the predator is grazing barnacle spat in preference to nearby adults (it is impossible to be certain of that from the photograph, but I removed the dog-whelk from the rock, to check). Very small mussels are under attack in fig. 55 and the edible winkles in fig. 58 were only about a quarter of the size of the dog-whelks feeding on them.

Some prey species may grow large enough to become immune from dog-whelk attack; mussels and perhaps some large species of balanid barnacles can manage this in Britain. In the Pacific, *Balanus cariosus* reaches size immunity after its first year of life by leaving cavities in its wall plates which render them too thick for *Nucella*'s accessory boring organ to penetrate. As its name suggests, *Perforatus perforatus*, found on southwestern shores of Britain and Ireland, has similarly hollowed wall plates. It grows appreciably larger than our mid-shore species, but I do not know whether it can become too large for *Nucella* to attack.

Life history

In *N. lapillus* the sexes are separate and fertilisation is internal. It is not easy to separate the sexes, although some authors have claimed that females have broader shells; the largest individuals in a population may all be females. When whelks are in aquaria, it may be possible to see the large penis behind the head of a male, on his right side, if he crawls up the side and females, crawling across the tank side, may be recognised through the glass. All *Nucella* have an indentation on the underside of the foot marking the position of the accessory boring organ; females have a second opening, posterior to that, marking the position of the pedal gland which is used to attach individual egg capsules to the substrate.

N. lapillus usually aggregates to spawn (fig. 60). Some thirty or more (sometimes very many more) individuals congregate in a moist shady place, often in a pool or cleft. In such a group some individuals may be observed copulating whilst others deposit egg capsules.

Fig. 61. Newly-hatched *N. lapillus* crawlaways feeding on *Elminius modestus*. The upper three arrows locate the dog-whelks; the lowest one, a mini-borehole into a young barnacle.

Females copulate repeatedly and lay a few capsules, one by one, between matings.

The number of capsules laid by a female is controlled by her food reserves, her size, her age and various environmental factors, including temperature. Each capsule contains around 600 eggs, 94% of which were laid as 'nurse eggs' for the nourishment of the remaining 6% which develop into embryos. The nurse eggs agglutinate to a mass in the centre of the capsular space. The embryos initially attach themselves to the mass and later move over its surface feeding.

Development of *N. lapillus* eggs is slow; in Somerset, adults lay egg capsules from late March to late April and hatching occurs over a two-month period commencing at the end of June. At least in some years, smaller numbers of apparently fresh capsules are to be seen in August. The equivalent of a veliger larval stage is completed within the capsule and the young snails escape to the outside

Fig 62. Dog-whelks from the same group of hatchlings, a few weeks later.

Fig 63. The theoretical structure of a stable population. The 0+ age-class will be the smallest and should be the most abundant (some will die in each subsequent year but there can never be any more).

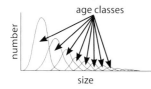

Fig 64. A length frequency plot for *N. lapillus* on Hurlstone Point, Somerset. April.

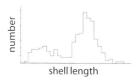

Fig 65. A young, actively growing dog-whelk has a shell with a thin sharp outer lip to the aperture.

world through a hole at the top as crawlaways (fig. 61). Observations of Somerset capsules suggested that between 12 and 15 crawlaways hatched from each capsule but there are records elsewhere of up to 35.

Recently-hatched dog-whelks may be found close to their capsules, preferring to hide in the empty cases of barnacles rather than to cling onto bare rock. It may be that their main requisite at this stage is to find protection whilst they continue to live off their food reserves remaining from the nurse eggs. When they start feeding they are said to prefer tiny mussels or spirorbid worms but some readily accept newly-settled barnacles. A few weeks later, the young dog-whelks were feeding on serpulid tubeworms and small *Balanus crenatus* barnacles (fig. 62).

Growth

The first-formed shell (the protoconch), borne by the crawlaway on emergence from the capsule, persists as the apex of the adult shell until such time as it is worn away. A snail increases the size of its shell by adding new material to the lip. As growth is never absolutely continuous, all shells show more or less well-marked growth lines, parallel to the shell lip, indicating periods when growth has slowed. Shells, like trees, bear for all time a record of the organism's growth pattern. If the growing lip is damaged it will be repaired, but the scar remains. As the animal grows it lays down thicker shell at the lip, whilst at the same time thickening the older sections from the inside. This latter is essential to strengthen the apex. Fretter and Graham (1962, 1994) give details of shell formation.

For species such as the common topshell, which (in Britain) has a short annual breeding season and in which individuals continue to grow throughout their lives, the structure of a population might be expected to resemble fig. 63 under stable conditions, individuals of successive year classes increasing in size whilst decreasing in numbers each year. In practice, field data never show this pattern because large individuals are more conspicuous and easier to collect than small ones.

A length frequency plot of *Nucella lapillus*, taken one April on Hurlstone Point (Somerset), is clearly revealing

Fig 66. On reaching sexual maturity, growth ceases and the outer lip of the shell is thickened and reinforced with 'teeth'.

Fig 67. A row of teeth inside the shell lip probably indicates a temporary cessation of growth during the winter before it reached sexual maturity.

Fig 68. Multiple rows of teeth inside the shell lip probably indicate that this individual had suffered parasitic castration through infection by the trematode *Parorchis acanthus*.

a different pattern. Most authors agree that growth in *N. lapillus* usually stops at maturity, approximately three years after the egg hatched. If this is true on Hurlstone Point, the lefthand peak on fig. 64 represents youngsters approaching their first birthday; the lefthand 'shoulder' to the main peak represents juveniles approaching their second birthday, and the remainder are adults.

Dog-whelks parasitised by cercaria larvae of the trematode fluke *Parorchis acanthus* never reach maturity and may continue to grow slowly throughout their lives.

Growth rate is related to the eventual size, individuals that will mature at a large size growing faster than those that stay small. Thus, whilst Feare's (1970) youngsters from Robin Hood's Bay were less than 10 mm at one year old, a little over 15 mm at two and entered adulthood at around 20 mm, some of my aquarium-raised progeny of large dog-whelks (around 45 mm adult length) had reached 30 mm by the end of their second year.

The shell of an actively-growing immature dog-whelk has a sharp lip (fig. 65). When the snail ceases growth, the lip is thickened and further armoured by a series of white lumps, commonly called 'teeth' (fig. 66). Should growth of an immature be checked temporarily, by prolonged winter cold or some accident, a similar line of 'teeth' will be formed and will remain visible on the inside surface of the shell wall when growth resumes (fig. 67). Shells bearing multiple rows of 'teeth' may be those of individuals parasitised by *Parorchis acanthus* (fig. 68).

Variation in shell size, shape and colour

There is a great deal of visible variation in shells of the common dog-whelk (plates 2 and 3). In this species, with no free-swimming stage in its life cycle, an individual's potential for dispersal away from the natal area is governed by its crawling activity. Feeding on an abundant sedentary prey, of which a year's supply may be obtained within one square metre, there is little stimulus for great activity; I have recovered marked individuals from within 30 cm of their release point a year later. It may even be that they return to breed in the same aggregation every

Fig 69. The extremes of adult size shown by *N. lapillus* in SW England. See also plate 2.

Fig 70. Spanish *N. lapillus* living in and under mussel clumps. Scale in cm.

Valdovino

Fig 71. The exposed shore morph of *N. lapillus* from an offshore Pembrokeshire island. Scale in cm.

Grassholm

year – capsules are certainly deposited in the same locations, year after year. So it would not be surprising to find that the species exists in more or less discrete genetic groups (sometimes called enclaves) even where, to a casual glance, there appears to be a continuous population stretching along miles of coastline.

Shell size must be influenced by the supply of food available to the juvenile snail but that cannot account for the size discrepancy in fig. 69 (or plate 2) which must involve a genetic component. The larger shell, from Gore Point in Somerset, is 56 mm long whilst the smaller one is from a sample collected at Hartland Quay in North Devon which had a mean adult shell length of 17.1 ± 1.4 mm. The largest *N. lapillus* shell that I know of was 62 mm, measured by Clive Roberts from members of a sub-littoral population washed ashore on ledges east of Kimmeridge Bay (Dorset) by a gale early in 1977. The mean size for the species is 27.5 mm.

Enclaves composed of very large individuals are usually sublittoral but may occasionally be uncovered on the extreme lower shore, as on Gore Point. The increased size and thickness may confer some protection from predation by large crabs and lobsters.

Shell colour and banding patterns (plates 2 and 3) are under genetic control and not obviously related to any environmental factors, despite some 1930s work which suggested that dark coloured shells had derived their pigment as a result of their owners feeding on mussels. Whilst it is true that mussels form part of the dog-whelk diet in some areas where distinctively coloured or banded individuals are common, such as north Cornwall, almost all dog-whelks are white in large areas of Shetland where mussels are also consumed regularly.

Banded coloured shells are undeniably cryptic to the human eye, especially in the case of small individuals that live under clumps of mussels (such as those in fig. 70 from North Spain) and perhaps they are also cryptic to birds. But bird predation is rarely a serious problem for dog-whelks. *N. lapillus* is distasteful to humans, even to sea-food enthusiasts, so perhaps it is for birds as well.

Fig 72. The sheltered shore morph of *N. lapillus* from deep within Salcombe Harbour, Devon. Scale in cm.

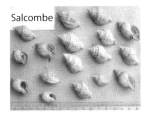

Salcombe

Fig 73. Exposed shore *N. lapillus* can fit into crevices.

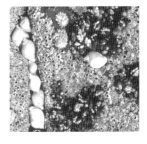

Fig 74. I maintain that the least ambiguous measurement of dog-whelk shell shape is the ratio *L/Ap*

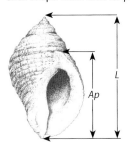

Shell shape is usually related to wave action and the intensity of crab predation. The most exposed shores are inhabited by dog-whelks with short squat shells that have a wide aperture (fig. 71); the most sheltered shores support snails whose shells have a long pointed spire and an elongated narrow aperture (fig. 72) whilst shores of intermediate exposure produce shells of intermediate shape.

Crabs are much commoner on sheltered shores than on exposed ones. The narrow aperture of the sheltered-shore dog-whelks (fig. 72) may prevent crab claws from obtaining a purchase on the shell lip. These snails do not, normally, occupy the upper reaches of the spire so that, *in extremis*, they can withdraw completely within their shell. At other times, they can hold more water within the shell; sheltered shores are drier than wave-washed headlands, at low tide.

Exposed-shore dog-whelks fit into crevices (fig. 73) and the wide aperture is required to accommodate the large foot which helps the snail cling to the rocks on exposed shores. Such a shape is, however, susceptible to crab predation but this may not be a problem as British crabs avoid high energy shores. The space inside the shell is almost entirely occupied by the body of the snail, when withdrawn, so there is no room for reserves of water - but desiccation is rarely a problem on shores that are well showered with spray even at low tide.

The ratio of shell length to aperture length, *L/Ap* (fig. 74), is a convenient measure of shell shape. *L* is the least ambiguous measurement than can be taken of a dog-whelk shell and, unlike breadth, is not dependent on orientation. *Ap* is taken from the point at which the suture ends; the point at which the outside of the outer lip of the aperture meets the body whorl. A sample of 30 adults, collected without conscious bias, is adequate to determine the mean (see p. 81).

The mean *L/Ap* ratio of the sample from Grassholm (fig. 71) is 1.25, whereas that from the Saltstone (fig. 72) is 1.47. In most parts of Britain, there is a strong correlation between mean *L/Ap* and the exposure grade, determined on Ballantine's (1961) scale (see p. 64). Changes in exposure round a headland are reflected in the mean shell shape of the dog-whelks found living

Fig. 75. The mean shell shape (*L/Ap*) of dog-whelk enclaves around the Garrison – the westernmost extremity of St Mary's, Isles of Scilly.

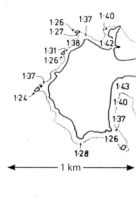

there. Fig. 75 depicts the situation around The Garrison, the most westerly point of St Mary's, Isles of Scilly. It is assumed that this pattern was caused and is maintained by natural selection.

However, this is not the pattern seen everywhere around our shores and fig. 76 displays the shell shape data I obtained along the English shore of the English Channel. At the western extreme, around the Isles of Scilly, where the rocky shores offer the whole range of exposure grades that Ballantine envisaged, the dog-whelks show the whole range of mean shell shapes from about 1.2 to 1.47. The same is true along the coast of mainland Cornwall and Devon, but east of the Exe estuary the mean values for *L/Ap* rise so that, east of Swanage, the value is always in excess of 1.4.

As mentioned above, *Nucella lapillus* is primarily a predator of prey living on hard rocks (although both barnacles and mussels can colonise stable shingle). Hard rock habitats are in short supply in the eastern half of the Channel, so dog-whelks were always commoner to the west but the disproportionate number of dots on fig. 76 west of the Lizard is also due to the effects of TBT pollution (see below).

Fig. 76. Variation in dog-whelk shell shape along the north coast of the English Channel. Mean shell shape, as measured by the ratio *L/Ap*, plotted against the eastings component of the Grid Reference. Data were collected mostly between 1966 and 1996. It will be noted that the exposed shore form is absent east of Swanage. The arrowed points refer to sub-littoral populations. The letters on the map identify the 100 km squares of the National Grid.

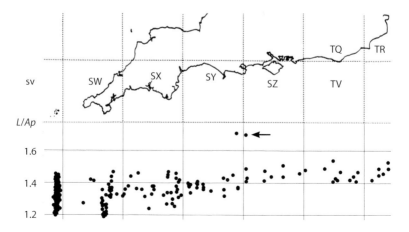

Fig. 77. Variation in dog-whelk shell shape with exposure to wave action.
Mean shell shape, as measured by the ratio *L/Ap*, is plotted against the exposure grade
on Ballantine's (1961) scale which ranged from 'extremely exposed' (grade 1) to 'extremely
sheltered' (grade 8).
(A) is the pattern I derived in 1973 from the area of Pembrokeshire used by Ballantine.
The slope of the regression line is given by y = 1.214 + 0.036x.
(B) is the equivalent pattern for the eastern part of the English Channel (less the sublittoral
data) with the same regression line to emphasise its inapplicability.

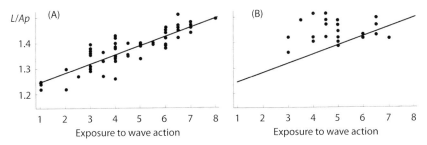

Fig. 78. The large form of *Nucella lapillus* from the extreme lower shore at Porlock Weir in Somerset.

It might be thought that the predominance of elongated dog-whelks in the eastern half of the Channel was simply a consequence of there being no exposed shores there but that is not the whole story. The relationship between shell shape and exposure that I derived from Pembrokeshire shores (fig. 77A) seems to apply on Atlantic shores from Brittany to Anglesey, SW Scotland and Orkney; but not in the Eastern Channel (fig. 77B) or the North Sea. There is a genetic component to shell shape, as well as that resulting from the selection acting on each new cohort of snails. It may be possible to demonstrate that such selection has occurred if juveniles are more variable than adults. An investigation of this kind would be best carried out in autumn when the current year's juveniles are big enough to measure but still distinguishable from the previous year's cohort and the adults.

Fig. 79 (overleaf) is another plot of mean dog-whelk shell shape against the eastings component of the grid reference, this time for the English coast of the Bristol Channel. As in fig. 76 the western end of the area shows the same range of variation as indicated in fig. 77A. Indeed, an earlier version of fig. 79, (Crothers, 1974), showed no value of *L/Ap* above 1.50 west of Gore Point.

Fig. 79. Variation in dog-whelk shell shape along the south coast of the Bristol Channel. The mean *L/Ap* ratio is plotted against the eastings component of the grid reference. Middle Hope marks the up-stream limit of *N. lapillus*. Data were collected between 1968 and 2001. The letters on the map identify the 100 km squares of the National Grid.

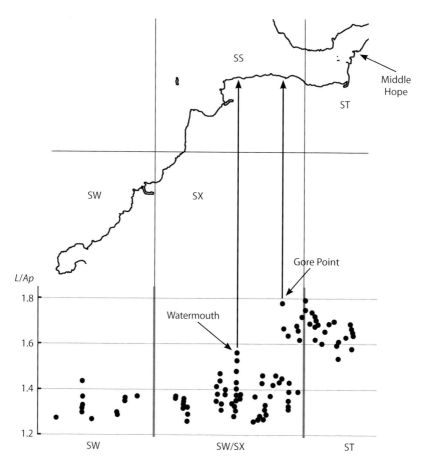

The sample of large dog-whelks from Gore Point (labelled in fig. 79) was collected at extreme low water mark and is illustrated in fig. 78. The largest shell also appears in fig. 69 and plate 2. These are shells from an effectively sub-littoral population; very similar shells wash up on some Dorset beaches.

Watermouth Cove is a beautifully sheltered inlet, just east of Ilfracombe. Ranked Ballantine Grade 7, it had a dog-whelk population with a mean *L/Ap* ratio of 1.47 in the 1970s. That population was eliminated by TBT pollution (see next section) and the site was eventually recolonised from a sub-littoral population. It will be interesting to see whether, in time, these elongated shells revert back to 1.47.

East of Gore Point, all enclaves are markedly elongated. This pattern has nothing to do with exposure to wave action; they breed true when reared in aquaria.

TBT pollution

During the 1970s, a series of new and improved antifouling paints became available for discouraging biological growth on the hulls of vessels. They incorporated tributyltin (TBT) and were intended to preserve the smooth contours of the hull as it passed through the water. As the new paint was comparatively expensive, it was disproportionately attractive to the owners of pleasure craft (who were more willing to afford the increased cost). The biocide was released slowly into the water, thus guaranteeing a longer-lasting deterrent than that provided by conventional paints. It was much more effective than anything that had been available before but, unfortunately, there were unacceptable side effects (collateral damage) that led to a ban on sale of the paint during the late 1980s. This was triggered by its effect on commercial shellfish, especially oysters, but it was equally serious for *N. lapillus*. At very low concentrations in sea water, TBT stimulated females to grow a penis which obstructed the egg canal and prevented egg-laying (Bryan, Gibbs, Hummerstone & Burt, 1986).

Recreational boating is most popular on south coasts, and *N. lapillus* was eliminated from most of southern England east of the Lizard (and, locally, in other parts of the UK – including Ilfracombe Harbour and Watermouth Cove). Currently, many sites are being recolonised. In the absence of a planktonic larva, it had been thought that *N. lapillus* had limited powers for achieving this! Are there more sub-littoral populations from which the shore can be repopulated? At least in some places, I think there are.

6 Winkles

Fig. 80. The shell of a flat winkle.

Fig. 81. An adult *Littorina fabalis* (left) and a juvenile *L. obtusata*.

Fig. 82. Penes of *Littorina fabalis* (left) and *L. obtusata* (from Reid (1996) which should be consulted regarding intraspecific variation).

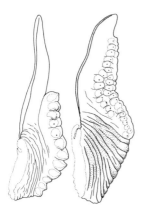

Winkle species

After limpets, winkles are *the* archetypal group of rocky shore snails; one or more species are found on almost all shores. It is easy to recognise that a snail is a winkle (a member of the family Littorinidae) but it was much more difficult to decide how many species were present (or which they were) before the publication of Reid (1996). A glance at his illustrations, showing the range of variation to be found within many of the species, explains the problems faced by earlier taxonomists. The winkle section of the key (pages 53–55 in this book) is based on David Reid's work and should make it *possible* (but not necessarily easy) to identify most individuals to species (or species pair). Certainty depends on characters that are only revealed by dissection. However, once it has been determined which species are present at a particular site, it may be comparatively simple to tell them apart using shell characters not mentioned in the key (because they would not apply elsewhere).

Flat winkles

On British and Irish shores, it is easy to distinguish flat winkles from the others, because of the low or very low spire to the shell (fig. 80), but care may be needed to separate *Littorina fabalis* from *L. obtusata*. Fig. 81 displays an adult *L. fabalis* (left), with a thickened shell lip, beside a juvenile *L. obtusata* of much the same size.

For certain identification it is necessary to examine the penis of an adult male (fig. 82). That of *L. obtusata* has 10–54 penial glands arranged in a double row and the terminal filament does not exceed 25% of the penis length. In *L. fabalis*, 3–14 penial glands are arranged in a single row and the penis ends in a long, worm-shaped filament that is 30–60% of the total length.

Food preference

Most *Patella* limpets, topshells and the other winkles are generalist grazers, scraping the algal felt off the rock surface in a wholly unselective way. By contrast, flat winkles

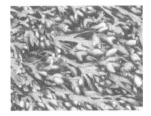

Fig. 83. Flat wrack, *Fucus spiralis*, upper shore.

Fig. 84. Bladder wrack, *Fucus vesiculosus*, middle shore.

Fig. 85. Egg wrack, *Ascophyllum nodosum*, middle shore.

Fig. 86. Saw wrack, *Fucus serratus*, lower shore.

feed on the stipes and fronds of the brown fucoid algae. (The blue-rayed limpet, *Patella pellucida* feeds on kelp and *Fucus serratus*.)

L. obtusata is thought to favour the middle-shore *Fucus vesiculosus* and *Ascophyllum nodosum*, whilst *L. fabalis* usually grazes the epiphytes off the lower-shore *Fucus serratus*.

Life cycle

In both species, the sexes are separate and fertilisation is internal. The female usually lays the fertilised eggs in jelly masses attached to her favoured food plant or, occasionally, on the rock surface underneath. As in dog-whelks, the equivalent of all larval stages is passed within the egg mass and the young snails hatch as crawlaways.

Egg masses attached to flat wrack, *Fucus spiralis* (fig. 83), bladder wrack, *Fucus vesiculosus* (fig. 84), or egg wrack, *Ascophyllum nodosum* (fig. 85) were probably laid by *Littorina obtusata*. Those on saw wrack, *Fucus serratus* (fig. 86), are probably those of *L. fabalis*.

Life expectancy of *L. fabalis* is about a year and that of *L. obtusata* is two or three years (Williams, 1990).

Colour variation

Both species show essentially the same range of colour variation, although the olive green morph (olivacea) is much commoner in *L. obtusata* and the yellow (citrina) morph in *L. fabalis*.

The polymorphism is under genetic control but the disproportionate abundance of particular morphs is presumed to be due to natural selection. Some authors have linked various morphs to environmental variables including salinity and temperature (both associated with height up the shore) but British workers have generally concentrated on predation as the selective instrument.

The olive-green morph of *Littorina obtusata* has a shell the size, shape and colour of the bladders on bladder wrack, *Fucus vesiculosus*, and is certainly cryptic to humans (fig. 87). Perhaps predators hunting by sight (birds and blennies) experience similar problems in spotting them and take a higher proportion of the more conspicuous individuals.

Fig. 87. I could see four *L. obtusata* in the frame when I took this photo - but only three now!

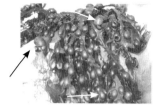

Fig. 88. Green algae growing over the yellow shell of *L. fabalis* and altering its colour.

Fig. 89. *Littorina compressa.*

The yellow morph of *L. obtusata* may be most abundant on flat wrack, *Fucus spiralis,* which has globular yellow receptacles at the end of the fronds in spring. A dark morph with net-like black markings (reticulata) has been linked to saw wrack, *Fucus serratus,* which has neither bladders nor rounded receptacles. The weed is dark brown during the summer but a good deal of black stipe is visible in winter.

As noted above, where both flat winkles occur together, *L. fabalis* is usually found on saw wrack, *Fucus serratus,* and is usually yellow. *F. serratus* is a lower-shore alga and so spends more time submerged than exposed; fish might, therefore, be more significant predators than birds. It turns out that pale yellow snails are almost invisible on *Fucus serratus* fronds under water, when viewed from below against the sunlight.

If the abundance of yellow *Littorina fabalis* is indeed due to its crypsis under water one might expect it to be most abundant in clear water, and least so in the muddy water at the mouth of an estuary.

A complication might arise if green algae, growing over the surface of a shell, altered its colour (fig. 88).

Rough winkles

There are three species of rough winkle. *Littorina compressa* is easily identified by its ridged shell, with the rounded spiral ridges wider than the grooves between them (fig. 89). The shell is usually yellowish green with black lines in the grooves. (This species was known as *L. nigrolineata* for several years but, alas, that name is not available as it had already been used for a black-lined variety of *L. saxatilis* in Spain.) The females lay their eggs in jelly masses attached to the rock surface and the young snails hatch as crawlaways.

The other two both show much the same range of variation, nationally, and were recognised as separate species on account of what the females do with their eggs. Eggs of *L. arcana* are laid in jelly on the rock but those of *L. saxatilis* are retained by the mother in a brood pouch within her mantle cavity. In both cases they hatch as crawlaways. Although there are no nationally applicable shell characters to distinguish the species, they can often

Fig. 90. Female *Littorina saxatilis* removed from their shells.

masses of purple embryos

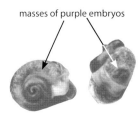

Fig. 91. Slight pitting in the concrete of Watchet harbour wall offers protection for small rough winkles, which have grazed away the algal felt.

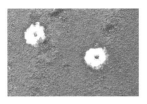

Fig. 92. Evidence of grazing by *Littorina saxatilis*. Granite is a pale-coloured rock which has been colonised by a dark algal film. Winkles, living in a crevice (top right) or amongst the seaweed (at the bottom) have grazed it away from the most accessible areas.

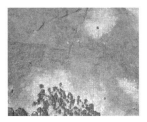

be separated without difficulty on those shores where both occur together. For example, on Hurlstone Point, Somerset, *L. saxatilis* has a ridged shell whereas that of *L. arcana* is smooth and has a notably wider aperture; a few miles east along the Somerset coast, at Watchet where *L. saxatilis* alone is present, it is smooth shelled.

To decide which was which, it was necessary to sacrifice a few of each. When they were dropped into boiling water and the body pulled out of its shell with a pin, female *L. saxatilis* were instantly recognised by the masses of purple embryos in their mantle cavity (fig. 90). (Adult females seem to carry embryos, at different stages of maturation, throughout the year.)

L. compressa is essentially a middle-shore species but the other two extend into the splash zone, above the upper barnacles and limpets. They overcome the same desiccation problems as do limpets, often clustering together in crevices (figs 91 and 92). They fill the space between their shell lip and the rock with mucus as the rock dries. This not only holds them in place but also hardens into a permeable membrane through which the snail can breathe but water cannot evaporate.

Although they do not make homes they may show homing behaviour. Those in fig. 91 appear to have grazed in a well-defined home range around a particular cranny. The white concrete is, elsewhere, invisible under a 'felt' of algae. A similar situation is seen in fig. 92, where the dark felt has been grazed in places to reveal the paler granite (which may be the explanation for the white border to the barnacle clumps in fig. 42). At low tide by day the winkles concerned are presumably amongst the algae or the barnacles. The pattern is rarely as obvious as this, but particular crevices and crannies always seem to have winkles in residence at low tide by day. Are they the same winkles each day? Marking the shells (and the 'home' crevice) in a distinctive way and checking them on subsequent days would answer the question.

As all three species lack a planktonic phase in their life cycle, the spread of genes would seem to be limited by the crawling activity of the snails. In some localities, shells of a distinctive colour are locally abundant, whilst

Fig. 93. The small morph of a rough winkle sheltering in an empty barnacle shell. (The barnacles are *Chthamalus montagui*, a species in which the wall plates fuse together in the adult.)

Fig. 94. *Melarhaphe neritoides.*

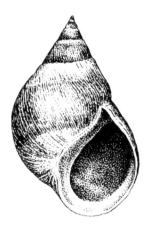

a short distance away a different colour stands out. It would be interesting to find out if these 'areas of colour dominance' are of similar sizes and/or whether their boundaries coincide with physical barriers of one kind or another.

Perhaps the most surprising variation shown by these animals is the range in size shown on different areas of the same shore. On barnacle-dominated shores, amongst empty barnacle shells on the middle shore, may be found examples of a small morph (fig. 93). (The winkles that caused the patterns illustrated in figs 91 and 92 were of this size or smaller.) They breed at this size and are often of a different colour morph from the, much larger, individuals on the upper shore. At one time the small morph was regarded as a separate species, *Littorina neglecta*, until it was realised that all three rough winkle species could be found represented by this small morph.

Littorina saxatilis (and probably *L. arcana* as well) shows considerable colour variation which appears to be cryptic in some cases; brown winkles are commoner on red rock and grey winkles on grey rock. This suggests the influence of predators that hunt by sight; rock pipits, *Anthus petrosus,* and herring gulls, *Larus argentatus*, are likely suspects.

The two remaining species of winkle, *Melarhaphe neritoides* and *Littorina littorea,* differ from the others in having planktonic eggs and larvae. It follows, therefore, that an individual female's progeny are distributed over a very much wider area and there has not been the same opportunity for the development of local varieties, forms and morphs.

Small winkles

Melarhaphe neritoides (fig. 94) is able to live even higher up into the splash zone than *Littorina saxatilis* with which species it overlaps at the top of the shore; they live so high that some individuals can never be submerged but only wetted by spray. Since the crawling stage (snail) has to settle from the swimming stage (veliger larva), settlement takes place towards the lower limit of their vertical range and individuals crawl upslope as they get older and are better able to withstand the desiccation experienced high up.

It may be that a size gradient exists, whereby the mean size increases with altitude. But there is also a sexual dimorphism whereby females grow larger than males; all individuals over 5 mm in length may be female. However, the females have to release their egg capsules (each containing a single fertilised egg) into the sea and, in order to do so, must descend below the high tide mark. Spawning is said to take place from September to April.

Unlike the other species of winkle, the periostracum (the horny outer layer of the shell) extends beyond the edge of the calcified shell lip providing a flexible rim to the shell aperture. The advantage conferred on the snail is that a better seal can be obtained against the rock surface to prevent desiccation of the animal inside.

Fig. 95. *Littorina littorea.*

Edible winkles

Littorina littorea (fig. 95) must be one of the best known of all rocky shore snails and the only one still gathered by the sackful (at least in Ireland) for sale as human food.

Edible winkles can reach very high densities. Although they graze the rock surface in a manner very similar to limpets and common topshells, they may also be found aggregated in hollows and pools where there does not seem to be any living algal growth (fig. 96). This winkle goes in for recycling in a big way. After the snail has scraped algae off the rock, the waste material that has passed through the alimentary canal is excreted as a series of faecal pellets. These provide a suitable substrate for bacteria, which further break down the faeces. Then along comes another winkle, which eats the faecal pellet, digests the bacteria and excretes fresh faecal pellets; these are colonised by bacteria and so on.

The young of this species spend a significant time at sea. The eggs, in floating capsules, take five or six days to hatch and the resulting veliger larvae may swim in the plankton for five or six weeks. After such a long period, the larvae may have been carried well away from their mother. This is the usual explanation for their scarcity (Isles of Scilly) or absence (Lundy) on offshore islands where the residual tidal drift (the net flow of

Fig. 96. An aggregation of *L. littorea* feeding on detritus; a combination of dead and decaying seaweeds and the bacteria living on faecal pellets.

water after tidal movements have cancelled themselves out) is from the open sea towards the mainland; no larvae arrive from other shores and most of those originating from adults on the island(s) are swept away.

Enclaves living on the coast of Somerset show a considerable range of population structure. On Hurlstone Point (a steep, hard rock, exposed shore, dominated by barnacles and limpets), there are few adults, but often a large numbers of juveniles, especially in the autumn. *L. littorea* is often extremely abundant on Gore Point (an extensive boulder shore on the opposite side of Porlock Bay) but individuals seem never to reach a large size. By contrast, the very much smaller enclave at Watchet (a very extensive, soft rock, wave-cut platform dominated by fucoid algae on the middle shore – see p. 64) is mainly composed of large individuals.

These discrepancies can be explained by differences in growth rate and survival, in response to the effects of exposure. On Hurlstone Point, conditions for settlement and juvenile growth are favourable in calm summer weather. Young winkles settle readily and grow well, but few survive the winter storms. Conditions are even more favourable for settlement on Gore Point and there the species attains a high density. The effects of wave action are less severe on this shore and there is little mortality or dislodgement during winter storms. Large numbers of individuals thus reach maturity and breed, but perhaps there is insufficient food to allow adults to both breed and grow. Growth is stunted or life is short. At Watchet, on the other hand, food is plentiful. The constraint here is the 'whiplash effect' of fucoid algae which sweeps away recently metamorphosed snails. Those that survive settlement grow rapidly to a large size and may live for many years. (The species is capable of living for 20 years, at least in an aquarium.)

7 Identification

Some 270 species of marine snail live in the seas around the British Isles (Graham, 1988) and their empty shells may be found on many beaches. Most are small (less than 3 mm long) or live beneath the tidemarks. The key, beginning on page 48, refers only to those species whose shells grow larger than 6 mm and which are regularly found *living* on rocky shores in some part of the British Isles. If a specimen does not key out easily, and match the accompanying illustration, reference should be made to Graham (1988)* or to Hayward & Ryland (1990, 1995).

This warning is particularly important here. A useful key could not possibly include every species that might, one day, be found on a rocky shore. The shore is continuous with the seabed and species from the sublittoral community occasionally stray up far enough to be found by people working on the shore at extreme low water of spring tides. Similarly, a journey around the shoreline will traverse sandy and muddy shores as well as rocky ones, and the tide may sometimes transport a hapless snail from one habitat to another. In the area of Somerset with which I am most familiar, the one additional species that I have twice recorded on rocky shores in a period of thirty years is the mud snail *Hydrobia ulvae* which lives on mudflats. Like numerous freshwater snails, it can hang from the surface film of a body of water. Under conditions of absolute calm, it may be transported tens of miles onto nearby rocky shores. An attempt to run this beast through the key would come to grief at couplet 26. Thus many species that are rarely found on rocky shores are strays from a vast pool of species more characteristic of other habitats.

That last sentence might raise a doubt about 'rarities' and it is worth considering what we mean when we use that word; "rarely encountered" or "numerically scarce". Species living close to the limits of their geographical distribution often fluctuate in occurrence, or abundance, from year to year. They may rarely be found at any particular place, but elsewhere may be very common.

---

* Alistair Graham's *Synopsis of the British Fauna* no 2 is out of print at the time of writing. It is currently under revision and will be reissued as two or more separate volumes.

Do rocky shores also harbour some species that are genuinely rare throughout their range? On land and in freshwater habitats, there are such species; very few individuals remain alive today. Often, this is because their favoured habitat has become fragmented and the survivors are effectively marooned on small islands of suitable habitat.

I wonder if that is possible amongst marine snails. Pelagic fish, mammals and reptiles range over huge distances and many snails and other invertebrates, benthic as adults, have a pelagic larval stage. In some species, viable larvae are thought to have crossed the Atlantic! Even snails that have lost this dispersal method are very widely distributed and are able to recolonise shores from which they have been temporarily eliminated by extreme weather or a pollution incident. Rex, Stuart, Etter & McClain (2010) discussed the problem in relation to a widespread yet generally rare deep-sea gastropod and concluded that there were probably a few breeding sites from which larvae were widely distributed.

Although some species complete the life cycle within a year, others may live for 15–20 years. As many individuals of these species continue to grow throughout their lives, shells may be found at any size from one or two millimetres up to the maximum noted in the key. However, as very few individuals ever reach their potential maximum size, it is unlikely that your specimen will be as big as the figure given.

Shells are frequently very attractive objects and beautifully coloured. However, colour variation within some species is so great (greater than the differences between similar species) that this is rarely a useful character for identification, which is why the key relies on line drawings.

There is no shore in the British Isles where all the species included in this key can be found alive so, in practice, recognition of the species you are looking at is much easier than the key might suggest. Readers who know that they are only going to work on the south coast of England may care to draw a line through all the species confined to northern Scotland, and *vice versa*.

Table 1. Name changes

Taxonomic revision, in many groups of rocky shore snails, has resulted in a significant number of scientific name changes in recent years. Natural history books are notably conservative in their usage of names, so this table is provided to help people match the snails identified with this key to information to be found in the literature.

Name used in the key	Name that may be found in the literature	Name that may be found in the literature	Name used in the key
Emarginula fissura	*Emarginula reticulata*	*Acmaea tessulata*	*Tectura tessulata*
Hinia incrassata	*Nassarius incrassatus*	*A. testudinalis*	*T. tessulata*
H. reticulata	*Nassarius reticulatus*	*A. virginea*	*T. virginea*
Littorina arcana	*L. saxatilis* (in part)	*Collisella tessulata*	*T. tessulata*
L. compressa	*L. nigrolineata*	*Cypraea monacha*	*Trivia monacha*
	L. saxatilis (in part)	*Emarginula reticulata*	*Emarginula fissura*
L. fabalis	*L. mariae*	*Helcion pellucida*	*Patella pellucida*
	L. littoralis (in part)	*Littorina littoralis*	*L. obtusata* and
L. obtusata	*L. littoralis* (in part)		*L. fabalis*
L. saxatilis	*L. neglecta, L. rudis*	*L. mariae*	*L. fabalis*
Melarhaphe neritoides	*L. neritoides*	*L. neglecta*	*L. saxatilis, L. arcana*
Nucella lapillus	*Purpura lapillus*		and *L. compressa*
	Thais lapillus	*L. neritoides*	*Melarhaphe neritoides*
Osilinus lineatus	*Monodonta lineata*	*L. nigrolineata*	*L. compressa*
	Trochocochlea lineata	*L. obtusata*	*L. obtusata* and
Patella depressa	*Patella intermedia*		*L. fabalis*
P. pellucida	*Helcion pellucida*	*L. rudis*	*L. saxatilis*
	Patina laevis	*L. saxatilis*	*L. saxatilis, L. arcana*
	Patina pellucida		and *L. compressa*
P. ulyssiponensis	*Patella aspera*	*Monodonta lineata*	*Osilinus lineatus*
Tectura tessulata	*Acmaea tessulata*	*Nassarius incrassatus*	*Hinia incrassata*
	Acmaea testudinalis	*N. reticulatus*	*H. reticulata*
	Collisella tessulata	*Patella aspera*	*P. ulyssiponensis*
T. virginea	*Acmaea virginea*	*P. intermedia*	*Patella depressa*
Tricolia pullus	*Phasianella pulla*	*Patina laevis*	*Patella pellucida*
Trivia monacha	*Cypraea monacha*	*P. pellucida*	*Patella pellucida*
		Phasianella pulla	*Tricolia pullus*
		Purpura lapillus	*Nucella lapillus*
		Thais lapillus	*Nucella lapillus*
		Trochocochlea lineata	*Osilinus lineatus*

Key to the larger species of marine snails that *live* on rocky shores around the British Isles

K1: The terminology of shell ornament.

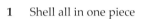

Suture: the continuous line running round the spiral of the shell, separating one whorl from another, and ending where the outer lip meets the body whorl.

Costa (plural, costae): numerous, often regular, ridges down the outer surface of the shell, running parallel to the outer lip.

Spiral ridges: numerous, often regular, ridges around the outer surface of the shell, running at right angles to the outer lip.

Varices (singular, varix): occasional pronounced costae, also called 'growth checks' (not obvious in K1, see fig. 31).

Body whorl: the last (largest) whorl of the shell, occupied by most of the animal's body.

K1   outer lip   aperture

1	Shell all in one piece	**2**

–	Shell composed of two, eight, or ten plates or valves	**not a snail**

Molluscs with the shell in two pieces are bivalves; those with the shell in eight pieces are chitons (coat-of-mail shells). Barnacles, which are not molluscs but sessile crustaceans, have a shell comprising a 'wall' of four or six plates and four moveable opercular plates in the middle.

K2

2	When alarmed, the snail withdraws *into* its shell	**3**

–	When alarmed, the snail withdraws *under* its shell	**30**

K3

3	On the top of its foot, towards the back end, the snail bears an operculum – a plate, often ear-shaped (K2) or circular (K3), used to close off the shell aperture when the animal withdraws inside	**4**

–	The snail lacks an operculum	**29**

4	Operculum calcareous – hard, white and shiny (K4). **Pheasant shell, *Tricolia pullus***

K4   operculum

–	Operculum chitinous, flexible, green, brown, orange or red	**5**

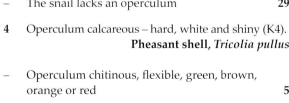

Plate 1.

1. The first-formed part of any snail's shell remains as the apex throughout its life, new material being added to the outer lip – or all around the margin in the case of limpets. In this case, it is easy to see the first year's growth distinct from that added in subsequent years.
2. Three painted topshells, *Calliostoma zizyphinum*, Hurlstone Point, Somerset.
3. Three common topshells, *Osilinus lineatus*, dried out, Garretstown Strand.
4. Another *O. lineatus* clearing a thin layer of sand off the bottom of a rockpool.
5. An elderly common topshell at Kilronan on Inishmore (Aran Islands) with growth checks suggesting that it had survived at least 15 winters.
6. A small breeding aggregation of dog-whelks, *Nucella lapillus*, Hurlstone Point, Somerset.

Plate 2. A selection of *Nucella lapillus* shells to display something of the variation in adult size and shape.

Plate 3. A selection of *Nucella lapillus* shells to display something of the variation in colour and banding to be seen at a single site on one day. Dunster Beach, West Somerset, in March 1985.

Plate 4. It is easy to separate adult common topshells, *Osilinus lineatus* (1), from purple topshells, *Gibbula umbilicalis* (2), on size, colour pattern and the presence or absence of an umbilicus. But when juveniles are involved complications arise, especially in the Channel Islands where there is a second species of purple topshell, *G. pennanti* (3). This photo was taken at Greve de Lecq, Jersey, in March.

Gibbula pennanti

Osilinus lineatus

Gibbula umbilicalis

5   At the front edge of the shell aperture (the lower end in these drawings) there is a notch, groove or tube through which the snail can extend a siphon (tube) to suck in water   Whelk **6**

—   The front edge of the shell aperture is a continuous smooth curve (as in K4)   **12**

6   The long siphonal groove is, at least partially, enclosed to form a tube (K5). Shell heavily ornamented

Sting winkle, *Ocenebra erinacea*

Lower shore, southern coasts. Up to 50 mm.

—   Siphonal notch or groove not forming a tube   **7**

K5   siphonal tube

7   In the living snail, the outer surface of the shell (the periostracum) is furry to touch. This 'fur' is rubbed off empty shells, when the curved nature of the costae becomes more apparent (K6). Shell with a siphonal notch rather than a groove   **Buckie** or

Common whelk, *Buccinum undatum*

Lower shore, all coasts. Up to 100 mm, but usually 15–40 mm on the shore.

—   Periostracum, if present, not brown and furry   **8**

curved costae

8   Shell may bear spiral ridges but not costae (K7), although one or two varices may be present

Dog-whelk, *Nucella lapillus*

Middle and lower shore, all coasts. Up to 60 mm, but usually less than 30 mm on the shore.

K6   siphonal notch

—   Shell bears costae and spiral ridges   **9**

9   Siphonal groove projects so far forwards (downwards in K1 and K5) as to make the shell appear pointed at both ends   **10**

—   Siphonal groove short, or reduced to a notch (as in K6)   **11**

K7   siphonal groove

K8

spiral ridge

K9

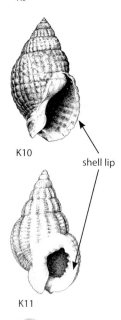

K10

shell lip

K11

K12

K13

10 Shell yellowish-grey or dirty grey with 16–19 fine spiral ridges on the body whorl (K8)

American oyster drill, *Urosalpinx cinerea*

Lower shore, Kent and Essex only. Up to 40 mm.

– Shell yellowish-white with 6–9 broad spiral ridges on the body whorl (K9)

young **Sting winkle,** *Ocenebra erinacea*

Lower shore, southern coasts. Up to 40 mm.

11 Costae and spiral ridges equally prominent, forming a pattern of squares over the surface. Shell with a siphonal notch (K10). Shell lip only thickened in adult specimens more than about 15 mm long

Netted dog-whelk, *Hinia reticulata*

Lower shore, all coasts. Up to 40 mm.

– Costae more prominent than spiral ridges. Shell with a short siphonal groove. Shell lip notably thickened and white (K11)

Thick-lipped dog-whelk, *Hinia incrassata*

Lower shore, all coasts. Up to 10 mm.

12 Operculum round (K12); inside of the shell nacreous (mother-of-pearl), white and shining

Topshell   **13**

– Operculum ear-shaped, or sub-oval (K13); inside of the shell dark or the same colour as the outside   **20**

13 Shell with an umbilicus (K14). (In such shells, the whorls do not touch in the centre of the spiral, leaving an open space; the umbilicus is the lower opening to that space)   **14**

– No open umbilicus, but its position may be marked by an 'umbilical scar'   **18**

14 A coloured layer covers the nacreous layer of the shell, more or less hiding it from view, except inside the aperture. Some of the coloured layer is frequently worn off, especially from the apex of the spire   **15**

umbilicus

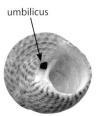

K14

– Shell white, glossy, faintly iridescent with greens or purples; lacking any superficial coloured layer so that the colours of the viscera show through (K15)

Pearly topshell, *Margarites helicinus*

Lower shore, northern coasts. Up to 10 mm.

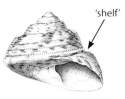

K15

15 Shell looks as though the spire has been squashed down into the body whorl, so there is a definite 'shelf' below the suture (K16). On this shelf there are various tubercles (bumps). Colours include red and cream but not green or grey

Turban topshell, *Gibbula magus*

Lower shore, south and west coasts. Up to 35 mm.

'shelf'

K16

– No obvious shelf visible below the suture and no tubercles on the shell surface. Colours include greens, greys and purple; rarely reds or cream **16**

16 A white lump (a 'tooth') projects into the aperture from the inner lip (K17). Decoration is a pattern of dark lines zigzagging over a green or light brown background (plates 1 and 4)

Common topshell, *Osilinus lineatus*

Upper middle shore, southwestern coasts. Up to 34 mm.

Very young common topshells have an open umbilicus, but in almost all British specimens it has closed by the time they reach 7 mm. Further south (for example in Portugal) half of the adult population may have an open umbilicus.

K17   'tooth'

– No tooth on the inner lip. Decoration composed of bands rather than lines **17**

17 Shell taller than broad (K18). Decoration a pattern of narrow dark grey or black zigzag bands on a pale grey background

Grey topshell or **Silver Tommy,** *Gibbula cineraria*

Lower shore, all coasts. Up to 18 mm.

K18

– Shell broader than tall (K19). Decoration a pattern of broad purple zigzag bands on a green background (plate 4)

Purple topshell, *Gibbula umbilicalis*

Middle shore, south and west coasts. Up to 22 mm.

K19

K20

18   Shell is an almost perfect cone (K20), with a very shallow suture. Either pure white or brightly coloured, purple, red, yellow or cream (plate 1.2)
Painted topshell, *Calliostoma zizyphinum*
Lower shore, all coasts. Up to 30 mm.

–   Outline of the cone interrupted by the suture   **19**

K21   'tooth'

19   A white lump ('tooth') projects into the aperture from the inner lip (K21). Decoration is a pattern of dark lines zigzagging over a green or light brown background (plates 1.3–5 and 4.1)
Common topshell, *Osilinus lineatus*
Upper middle shore, southwestern coasts. Up to 34 mm.

–   No tooth on the inner lip. Decoration is essentially a pattern of red or purple zigzag bands on a greyish-green background (K22, plate 4.3)
Gibbula pennanti
Middle shore, Channel Islands. Up to 18 mm.

K22

Very similar to *Gibbula umbilicalis* but without an umbilicus, and so may be distinguished from that species, without great difficulty, on shores where they both occur.

K23

20   The buff-coloured horny outer layer of the shell (the periostracum) is heavily ridged and conspicuous. Suture very prominent (K23)
Thick chink shell, *Lacuna crassior*
Lower shore, most coasts, rare. Up to 14 mm.

–   Periostracum inconspicuous or absent   **21**

K24   umbilicus (chink)

21   Shell with an umbilicus (K24), or chink. (In such shells, the whorls do not touch in the centre of the spiral, leaving an open space; the umbilicus is the lower opening to that space)   Chink shell **22**

–   There is no umbilicus. The inner margins of the whorls are in contact and form a solid rod up the centre of the shell (the columella)
Winkle **24**

K25

22 Shell with a high spire (K24), buff-coloured with three or four orange bands running round the whorl at right angles to the lip

Banded chink shell,
Lacuna vincta

Lower shore, all coasts, on red seaweeds. Up to 10 mm.

– Shell with a low spire, not banded **23**

K26

23 Aperture enormous, appearing to occupy more then half of the total area of the shell; spire low, virtually non-existent (K25). Pale green

Pallid chink shell,
Lacuna pallidula

Lower shore, all coasts, on *Fucus serratus*. Males up to 6 mm, females up to 12 mm.

K27

– Aperture appearing to occupy much less than half the total area of the shell which has a distinct spire (K26). Buff-coloured

Small chink shell,
Lacuna parva

Lower shore, all coasts, on red seaweeds. Up to 4 mm.

K28

24 Periostracum extending beyond the lip of the calcified layers, appearing as a colourless flexible rim to the shell lip. Shell long and pointed with a shallow suture (K27); the colour of the bloom on purple grapes

Small winkle,
Melarhaphe neritoides

Splash zone, all coasts except southeast England. Up to 9 mm.

– Shell lip without a colourless flexible rim **25**

25 Shell with a distinct spire **26**

– Shell spire low or very low (K28)   Flat winkle **28**

K29

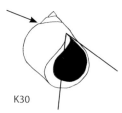

K30

26 **Either** shell more than 26 mm long or shell with a series of dark lines crossing the shell outer lip at right angles (K29). Suture shallow (arrowed in K30). Outer lip of shell meets the body whorl at an acute angle (K30). Black markings run *across* the head tentacles

Edible winkle, *Littorina littorea*
Middle and lower shore, all coasts. Up to 56 mm.

– Shell length 26 mm or smaller. Black lines cross the shell lip *only* if the shell has a black line in each spiral groove. Suture deep (arrowed in K31). Outer lip of shell meets body whorl at an obtuse angle (K31). A black line runs *along* either side of the head tentacles Rough winkle **27**

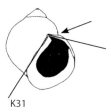

K31

27 Spiral ridges of the shell ornament have rounded tops and are wider than the grooves between them. Often the grooves are black, contrasting with the yellow or greenish ridges (K32)

Littorina compressa
Middle shore, north and west coasts. Up to 26 mm.

K32

– Shell smooth (K33) or, if ridged, with sharp-crested ridges narrower than the grooves between them
Either ***Littorina saxatilis***
Upper middle shore to splash zone, all coasts. Up to 26 mm.

or ***Littorina arcana***
Upper middle shore to splash zone, all coasts except southeast England. Up to 17 mm.

K33

The essential difference between these two species is that whilst females of *L. arcana* lay their eggs in masses of jelly attached to the rock, those of *L. saxatilis* retain them within their shell and appear to give birth to live young. On any particular shore it is usually possible to separate the two species (where both occur together), but there are no reliable characters applicable nationally, as both species display similar ranges of variation. *L. saxatilis* is the more widely distributed and much the commoner in sheltered sites. See Reid (1996).

K34

K35

K36

28 Found on or near flat wrack (*Fucus spiralis*), bladder wrack (*Fucus vesiculosus*) or egg wrack (*Ascophyllum nodosum*). The population contains animals of at least two markedly different sizes (separate generations). The shell has a detectable spire (K34); if it is smaller than 12 mm it probably has a thin sharp lip

   Common flat winkle, *Littorina obtusata*

   Middle shore, all coasts. Up to 20 mm.

— Found on or near saw wrack (*Fucus serratus*). All members of the population belong to the same generation (but there is some sexual dimorphism). The shell has no detectable spire (K35). Shells between 9 and 12 mm usually have a thick lip

   Annual flat winkle, *Littorina fabalis*

   Lower shore, all coasts. Up to 12 mm.

   The two species of flat winkle are not easy to separate. Where they both occur together, *L. obtusata* grows to a larger size. Both show the same range of colour variation, but flat winkles with olive green shells are most likely to be *L. obtusata* (see fig. 81, p. 38)

K37

29 Shell aperture a long narrow slit (K36)

   Cowrie, *Trivia monacha*

   Lower shore, all coasts. Up to 12 mm.

— Shell aperture occupies the whole under-surface of the shell   **30**

K38

30 Shell with perforations (a slit, hole or holes) in its surface in addition to the aperture (K37–39)   **31**

— The aperture is the only opening in the shell   **33**

K39

31 The perforation is a slit (K37)

   Slit limpet, *Emarginula fissura*

— The perforation is a hole or holes   **32**

K38

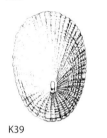

K39

K40

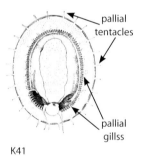

pallial
tentacles

pallial
gillss

K41

K42

32 A line of holes mark where exit siphons expel
water from the mantle cavity (K38)

Ormer, *Haliotis tuberculata*

Lower shore, Channel Islands. Up to 90 mm.

As the animal grows, and the mantle cavity moves forward,
redundant holes are sealed from inside.

– A single hole at the apex of the shell (K39)

Key-hole limpet, *Diodora graeca*

Lower shore, all coasts except North Sea. Up to 25 mm.

33 Shell a cone with strong ribs radiating out
from the apex (K40) *Patella* limpet **34**

– Shell without strong *radiating* ribs. **36**

To identify members of the genus *Patella* to species, it is
necessary to take the limpet off the rock – which will almost
certainly injure or kill it. Hence, unless it is essential to
determine the species, try to use such characters as *are*
visible. It is often possible to see the pallial tentacles
(K41) on limpets crawling around in rock pools.
  The apex of the shell is over the head, so the head is to
the right in K40. The other limpet shells on this page are
shown with the head downward.

34 Hind margin of the shell smoothly rounded
(K42). Pallial tentacles colourless. Sole of the
foot dusky grey

Common limpet, *Patella vulgata*

Whole shore above lowest levels, all coasts.
Up to 60 mm.

– Hind margin of the shell angulated. Pallial
tentacles white or cream coloured **35**

35 Sole of foot apricot in colour. Shell a very low
cone decorated with a large number of rather
fine radiating ribs (K43)

China limpet, *Patella ulyssiponensis*

Lower shore (and rock pools), all coasts except eastern
England. Up to 50 mm.

K43

– Sole of foot black. Shell usually a low cone decorated with heavy radiating ribs (K44)
Black-footed limpet, *Patella depressa*
Middle shore, south and west coasts. Up to 30 mm.

36 Shell a nearly symmetrical cone bearing an ornament of concentric ridges, each parallel to the shell margin (K45)
Chinaman's hat limpet, *Calyptraea chinensis*
Lower shore, south west coasts. Up to 15 mm.

K44

- Shell not a symmetrical cone (the apex is nearer the front end); shell surface smooth, with a pattern of radiating lines. No concentric ridges   37

37 A lower-shore limpet living on kelp (*Laminaria*) or saw wrack (*Fucus serratus*). The brown shell bears radiating flashes of iridescent blue (K46). The animal has pallial gills (secondary gills lying in the pallial groove, around the foot, K41)
Blue-rayed limpet, *Patella pellucida*
Lower shore, Atlantic coasts. Up to 18 mm.

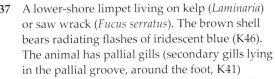

– A lower-shore limpet living on the rock surface. Shell without blue flashes; no pallial gills   38

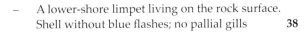

38 Shell basically cream or greenish marked by chocolate brown splodges (K47)
Tortoiseshell limpet, *Tectura tessulata*
Lower shore, northern and western coasts. Up to 20 mm.

K45

– Shell basically white or yellowish marked by pink lines, often broken into splodges (K48)
White tortoiseshell limpet, *Tectura virginea*
Lower shore, most coasts. Up to 10 mm.

K46

K47

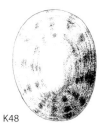

K48

8 The rocky shore environment

Tides

More than any other physical factor, tides define the sea shore. Should you have access to a harbour wall to which someone has fixed a depth guage on which you can record the height of the water every hour, and should you plot your data on a graph they might resemble those in fig. 97. Alternatively, you could derive similar figures from the *Admiralty Tide Tables Vol. 1 European Waters*, which is what I did. The vertical scale is in metres whilst the horizontal scale is in hours, from midnight to midnight.

Around the coasts of Britain and Ireland, we experience two high and two low tides, of approximately equal amplitude, during each period of (slightly more than) 24 hours. Especially in the vicinity of islands, the tidal pattern may be more complicated than in this example; in a few areas, including the central section of the south coast of England, the patterns are complicated by a high water 'stand'.

Comparing fig. 97A and B, we see that there is substantial variation not only in the times of high and low tide but also in the tidal amplitude. If we now look at all the predictions of high and low tide for a single month, for September in the case of fig. 98, another pattern

Fig. 97. Tidal predictions for the Port of Avonmouth for (a) February 28th and (B) June 3rd 1975. In each case the dots represent the predicted height of the water on the hour every hour from midnight to midnight (from Crothers, 1976). In this and subsequent figures MHWST indicates mean high water level of spring tides; MHWNT, mean high water of neap tides; MLWNT, mean low water of neap tides; and MLWST mean low water of spring tides. EHWST and ELWST indicate the extreme high and low water levels of spring tides.

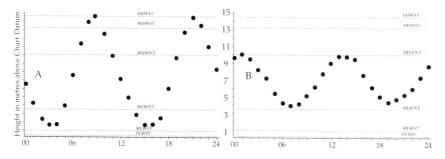

Fig. 98. Predicted heights of high (upper row) and low (lower row) tides at Avonmouth in September 1974 (from Crothers, 1976). mtl indicates mean tide level, the average of the tidal predictions.

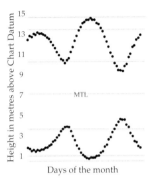

Days of the month

emerges. The tidal amplitude expands and contracts at approximately weekly intervals so that we get two periods of large amplitude tides (springs) and two periods of small amplitude tides (neaps) every *lunar* month; so there are 26 periods of spring tides and 26 periods of neaps in every calendar year. (Note that the word 'spring', when used in this context, has nothing to do with a season; it has more to do with 'springing up'.) Spring tides occur slightly after full moon and new moon, neaps at first quarter and last quarter of the lunar cycle.

The equivalent graph of the predictions for a calendar year (fig. 99) shows that the greatest variability in tidal amplitude occurs in spring and autumn (at the equinoxes) and the least in mid-summer and mid-winter (at the solstices). As in fig. 98, the upper line of points represents high tide levels and the lower line, low tide levels. The area above MHWST is called the splash zone and that below MLWST, the sub-littoral fringe.

From the data in fig. 98, it is possible to deduce how often the various levels on the shore are inundated with sea water (fig. 100A). The result is hardly rocket science;

Fig. 99. Predicted heights of high and low water throughout 1983 at Avonmouth. Spring tides - higher high tides and lower low tides - are seen to alternate with neaps throughout the year. The upper shore lies between the Mean High Water levels; the middle shore between the Mean Neap tide levels and the lower shore between the Mean Low Tide levels.

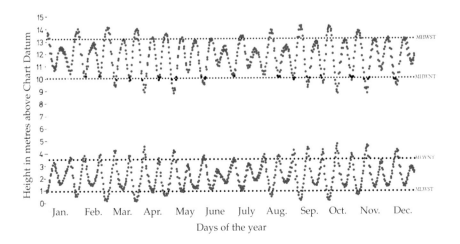

Fig. 100. Variation in the widths of the shaded areas represents the differential effects of the three dominant physical factors at different heights up and down a sea shore, calculated from the data used to plot fig. 99. A. inundation by sea water; B. the availability of light and C. the frequency of experiencing wave action.

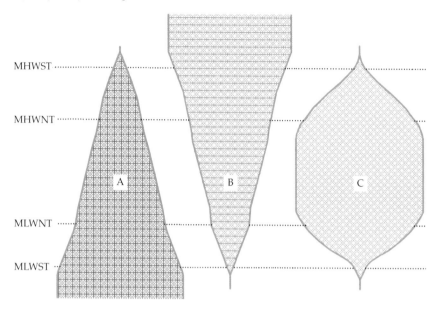

Fig. 101. The combined effect of the three dominant physical factors identified in fig. 100.

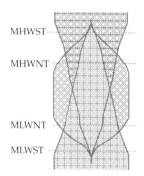

all it shows is that the rock is essentially dry above MHWST and is wetted more and more often downshore until below MHWST it is under water almost all the time. A wide range of effects are distributed across the shore in consequence. The rock experiences a far wider range of temperature when dry than when under water and the salinity can range far above and below that normal for sea water, from fresh water to crystalline salt.

Less obvious is the availability of light across the shore (fig. 100B). In estuaries, the sea bears such a huge sediment load that obstruction of light is inevitable; when paddling, your toes have disappeared from view before your ankles get wet! But even on oceanic coasts, with clear water, light does not pass through sea water as easily as it does through the atmosphere. Much of the incident radiation is reflected off the surface, especially early in the morning, late in the evening and in winter. And much

Fig. 102. Variation in the mean tidal range at ports around Britain and Ireland.

Barra 3.6 m

Oban 3.3 m

Greenock 3.0 m

Port Ellen 0.6 m

Ballycastle 1.0 m

Londonderry 2.3 m

Machrihanish 0.5 m

Avonmouth 12.3 m

Swansea 8.6 m

Milford Haven 6.3 m

Plymouth 4.7 m

Portland 1.9 m

of what does pass through the surface is absorbed, especially at the red end of the spectrum – that most useful for photosynthesis.

If these were the only two factors affecting seaweed distribution we should expect maximum growth on the middle shore, where the availability of neither water nor of light is limiting. In practice, it is often on the middle shore that fewest macroscopic seaweeds are present, due in part to the effects of a third physical factor, wave action. Breaking waves are a feature of the edge of the sea. The amplitude (size) of the waves is controlled by the strength of the wind, the distance that wind has been blowing over the water (the fetch) and the depth of the water; waves break when the depth falls below a third of the wavelength. The largest waves are probably experienced on the upper shore, but never for very long. Above MHWNT and below MLWNT, there are many days each month when the edge of the sea never reaches that level (fig. 99) but the middle shore, between the neap tidemarks, experiences the line of breakers four times every day. So the greatest frequency of wave action is experienced on the middle shore (fig. 100C).

Members of the rocky shore community are influenced by the combination of these three physical factors (fig. 101) and it is a harsh environment in which to live. At the top it is too dry (and often too hot or too cold), at the bottom it is too dark, whilst in the middle you get thumped by waves. To survive at all, rocky shore species have had to adapt to a particular combination of the factors. That is why the phenomenon of zonation is shown so clearly on some shores.

All of these graphs were compiled using the data for Avonmouth, the standard port with the largest tidal range of any around Britain and Ireland. Details of the pattern show up more clearly in these data than they would for some other places. The bars on fig. 102 indicate mean spring tidal amplitude for all the standard ports; the figures range from 12.3 m at Avonmouth to 0.6 m at Port Ellen on Islay. Atmospheric pressure may have more influence on water level than has the tide here, and locals talk about the tide being "out this week".

Machrihanish, on the west coast of Kintyre, has the smallest tidal range (0.5 m). As this site (it is hardly a 'port' by normal standards) is open to the Atlantic, tides are barely discernible (as I experienced on one occasion whilst 'waiting for the tide to go out'; it didn't).

Residual tidal drift

The water that floods across the shore on the rising tide is mostly that which drained away on the previous ebb. Until this was appreciated, it had been normal practice to discharge raw sewage into the sea (from small coastal settlements) on the ebb tide. It was often possible to see the plume carrying the waste out to sea (because the fresh water discharge floated on the sea water). When 'solids' were found on the beach at low tide, accusing fingers were generally pointed at the nearby settlements 'up tide'. I know of a Local Authorities dispute, in the late 1960s/ early 1970s, that led to an accused Rural District Council secretly releasing markers (from a boat, at night) into the accusing Council's sewage plume. They managed to demonstrate that most if not all of the 'solids' discovered on the accusing Council's beach had originated from its own sewer!

Thus, free floating eggs and larvae, released into the sea during one period of high tide and carried away on the ebb, will be brought back to much the same place on the following high tide. But in some (perhaps many) parts of the British and Irish coasts there is a net movement of water in one direction. The longer the eggs and larvae spend in the water, the further they will be carried from their parents.

Around the Isles of Scilly, for example, the residual tidal drift is towards the Cornish mainland. Common topshells, whose sea-going stages are completed within four days, are abundant on the sheltered and moderately exposed shores of those islands. Edible winkles, on the other hand, are rare on Scilly as, by the end of their six or seven weeks of pelagic life, residual drift will have carried most of the offspring from any Scillonian parents well away from the islands and will have prevented the offspring of Cornish mainland populations from reaching them. Crisp & Southward (1958) discuss this matter in more detail.

The dog-whelk *Nucella lapillus*, lacking a planktonic phase in its life cycle, is not directly influenced by residual drift; only indirectly through its influence on the availability of prey species. The barnacle *Semibalanus balanoides* and mussel *Mytilus edulis*, favoured prey elsewhere, are rare or absent on Scilly – presumably because their larvae are also transported away from the islands – forcing the predator to develop a different feeding strategy (see p. 27).

In both the English and Bristol Channels, the topshells *Gibbula umbilicalis* and *Osilinus lineatus* extend further east along the southern shore than along the northern one. Both species, here, are at the limit of their geographical range but it is probably not temperature that has restricted that eastward limit on the northern shore. (South-facing coasts are usually warmer than north-facing ones.) The explanation is more likely to be that the residual drift carries larvae from Biscay around Ushant and then eastwards along the north Coast of France. On the English side, residual drift may be westward, northward around Land's End and then eastward along the English coast of the Bristol Channel. Residual drift may be westward along the Welsh coast. Larvae from North Devon and Somerset populations are carried up into the low-salinity and turbid water of the Severn estuary and do not survive to colonise the Welsh coast.

Fig. 103. Two Scillonian shores, approximately 5 km apart, and both formed from the same granite rock.
A. Peninnis Outer Head, St Mary's, open to the southwest.
B. The east coast of Tresco, sheltered from the southwest.

Exposure to wave action

Although the amplitude and timing of the tides are usually similar on neighbouring shores, the resulting fauna and flora may be notably different. Setting aside the possible effects of human activities, the two main reasons for this are exposure to wave action (fig. 103) and geology.

Shores experiencing a long fetch can expect to receive the full effects of wave action. The longest fetches experienced on British and Irish shores are on west coasts, open to the Atlantic and not protected by offshore islands or reefs (fig. 103A, cover).

Not surprisingly, soft rocks erode more rapidly than hard ones. In the time available since sea levels rose to approximately their present position after the last Ice Age, many hard rock shores have eroded little and still present

Fig. 104. The extensive wave-cut platform of the shore east of Watchet Harbour, Somerset, showing changes in the cover of fucoid algae over the years.

1972

1977

1981

1995

1999

steep profiles. Water depth close to the rock surface may be considerable and the waves may break directly onto the shore. On open coasts, such shores are exposed to wave action.

By contrast, soft rocks have been eroded down to a wave-cut platform, presenting a low profile (fig. 104). The sea is shallow as the water's edge covers and uncovers the platform; any large waves break well to seaward and by the time they pass over the rock, it is cushioned by metres of water. In effect, such shores are sheltered from wave action.

The effects of wave action are often obvious, but measurement is fraught with difficulty. Attempts to record pressure or other impulses onto the rock have been thwarted because instruments capable of recording small changes are destroyed by large ones. Further, it is not the waves now breaking on the rocks that have affected shore life, but the effect of their predecessors months, years, decades and longer ago.

As a PhD student working, during his vacations, at Dale Fort Field Centre in the 1950s, Bill Ballantine (later Director of Otago Marine Laboratory in New Zealand) devised a biological exposure scale which enabled people to assess exposure from its effects instead of its causes. His benchmark paper included the figures reproduced here as fig. 105. Fig. 105B depicts a shore exposed to moderate wave action; fucoid algae are confined to the lower shore (when this was published, the barnacle *Semibalanus balanoides* was included in the genus *Balanus*). Fig. 105C shows a shore sheltered from wave action and dominated by fucoid seaweeds (much as fig. 104 in 1972 and 1999).

Perceived wisdom, at the time, deduced that wave action prevented fucoids from colonising the middle shore on exposed sites. But extremely exposed shores, fig. 105A, *do* support some middle-shore fucoids (*Fucus vesiculosus* forma *evesiculosus* is the bladderless bladder wrack).

It is the effect of exposure on *Patella* limpets that accounts for the differences seen in fig. 105. As mentioned earlier (p. 11), algae grow where they have not been eaten and the favoured habitat of *Patella* limpets is on moderately exposed shores. Where limpets have been removed from such shores, either experimentally (e.g. Jones, 1948)

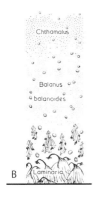

or following an oil spill, algae colonise and temporarily dominate the site, which reverts to its original condition when the *Patella* population has recovered.

The same species of fucoid algae live on the Atlantic shores of North America, but there it is the exposed shores that are seaweed dominated whereas the sheltered ones are mostly bare rock! Ice forms on sheltered shores at low tide in winter and strips the rocks bare as it is lifted off by the rising tide. There are no *Patella* on the eastern seaboard of the USA or Canada to control the fucoids on exposed headlands.

When first published, Ballantine's exposure scale was criticised on two counts: first because it was a tautology (he used the community to classify the shore and then used his classification to interpret the community), and secondly because it was not clear what, if anything, was being measured. Nevertheless, it has proved useful in many circumstances.

It must always be remembered that it is a biological scale, grading sites according to the differential responses of the shore fauna and flora to changes in exposure. The fauna and flora also respond to other environmental factors, especially temperature. Fig. 104 shows how a shore may change over time; the exposure to wave action remained the same but the relative abundance of limpets and of fucoids did not.

I interpret this sequence as follows. In 1972, as in all years between 1967 and 1976, the dominant fucoid cover inhibited the settlement of limpet spat. In 1975 and 1976 the long hot summers inhibited the settlement (or subsequent survival) of the fucoid algae. So in 1977, when the 1974 year class died (most fucoids seem to live for three years) there was nothing to prevent the arrival of baby limpets. These were, however, too small or at too low a density to prevent the recolonisation by fucoids. So in 1981 full fucoid cover was restored. But under the seaweeds there were large numbers of large limpets which seriously impeded the survival of the next generation of fucoids. And the fucoid/limpet balance continued to oscillate thereafter.

Fig. 105. Three pictograms from Ballantine (1961). (A) An extremely exposed, (B) a moderately-exposed shore (the *Chthamalus* is *C. montagui*) and (C) a sheltered rocky shore.

9 Techniques and approaches to original work

Planning an investigation

A clear theme underlying all of the *Naturalists' Handbooks* is an intention to encourage you, the reader, to contribute to the sum of human knowledge by undertaking fresh investigations and communicating the results to other naturalists. Communicating the results is important, whether the investigation is to be published or to be included in a submission for some award or qualification. In the latter case, assessment will be based on the impression made by the communication of the results.

A popular concept of science may imagine the inspired researcher collecting data on a whim, and it has happened; David Lack published observations of bird migration through an alpine pass made when his car broke down and he was awaiting help. But it is generally accepted that the best results are obtained through the discipline of hypothesis testing. This involves making an observation, proposing an hypothesis to explain that observation, making predictions on the basis of that hypothesis, and then devising an investigation, collecting and analysing data to see whether or not those predictions have been fulfilled. If the predictions are fulfilled, your hypothesis may be one valid explanation – but not, necessarily, the only one and you cannot accept it with certainty. On the other hand, if the predictions are not fulfilled, you can confidently reject your hypothesis. By convention, therefore, it is usual to devise a null hypothesis and design an investigation to see whether it can be rejected with confidence. For example, if your hypothesis is that dog-whelks are selecting their prey by size, your null hypothesis (abbreviated to H_0) might be that the size-frequency distribution of items taken as prey is not significantly different from the size-frequency distribution of potential prey items in the environment. If you can show that this is not the case, your hypothesis may hold.

When planning an investigation it is helpful to think things through in this sequence, but it is advisable to work backwards as well. Should your investigation yield the expected result, how will you be able to convince other people that the result is meaningful? It will probably be necessary to apply a statistical test to demonstrate that the result was unlikely to have been obtained by chance.

It is unwise to collect data first and then look for an appropriate test. Instead, you are recommended to select a suitable test and plan to collect data in a manner that will render application of that test straightforward. For example, if the rubric says the test is suitable for between 7 and 30 pairs of measurements, plan the investigation to deliver accordingly; somewhere in the low twenties would appear advisable in this case. I admit bias, in that I was the editor, but the *Open University project guide* by Chalmers & Parker (1989) will prove very helpful, as will *Studying invertebrates* (Naturalists' Handbook no. 28) by Wheater and Cook.

Most biologists find it easier to understand a graphical presentation than columns of figures so consideration should also be given to the manner in which the result will best be presented. Data should be collected accordingly.

Over the years, I have been involved in many field investigations, ranging from GCSE projects to PhD theses and longer term investigations carried out by my staff. Many of the most successful ones were devised by younger students; some of the most disappointing, by undergraduates. It is not only military planners who should remember the first Duke of Wellington's adage: "time spent on reconnaissance is seldom wasted." Success is closely related to the amount of time and effort devoted to planning – including, if at all possible, the collection of a pilot set of data. Problems have resulted from an inability to find or to identify the chosen species, an inability to sample in the planned manner at the chosen site and, most often, a gross underestimation of the time required.

Transects

One of the most obvious features of a rocky shore is the well-marked vertical sequence of different groups of species. From fig. 101 it is clear that the effects of the physical factors vary across the shore in a pattern that

Fig. 106. Zonation of rocky shore snails just west of Gore Point, Somerset, in 1974.

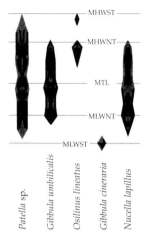

Fig. 107. The use of a cross staff when sampling along an interrupted belt transect.

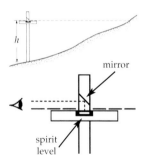

is related to height above Chart Datum (the basal level chosen by the hydrographer who drew the relevant chart). So it is logical to record the distribution of the species against the vertical scale. This is (theoretically) easy up a harbour wall or a vertical cliff, but more challenging on broken shores that experience a large tidal range, such as that illustrated in fig. 104.

If we are to interpret this zonation of species on rocky shores in terms of their different adaptations to aspects of the shore environment, we need a way of recording and quantifying their distribution patterns. One way to do this is to record the species present at different levels using a transect up the shore from low tide level to MHWST or beyond. A transect is, literally, a cross section and is the classical technique for investigating a linear change across a habitat; transects have been extensively employed on rocky sea shores (for example, fig. 106).

Line transects, in which we record every individual that happens to be touched by a single thread run down the shore, are rarely applicable for snails. **Belt transects**, in which we record all individuals living within the confines of a strip of specified width, may be useful on near vertical sites with a tidal range of 2 m or less. But for assessments on more extensive shores, or when comparing shores with different tidal ranges, an **interrupted belt transect** appears most appropriate; the belt is assumed to run all the way up the shore, but information is recorded only at specified intervals.

The Oil Pollution Research Unit, managed by Field Studies Council from 1967 until 1995, recommended taking sampling 'stations' at vertical intervals (h in fig. 107) equal to (approximately) one tenth of the tidal range, so that ten stations were sampled on every shore. To determine the position of the stations, OPRU scientists devised a cross staff (fig. 107). A spirit level was set into the cross piece and a mirror inserted into the upright above it at an angle of 45°. The user squints along the top of the cross piece, checking that the bubble (seen through the mirror) is central in the spirit level, and determines the next point on the rock surface that is exactly h cm above the base of the cross.

Fig. 108. Using a length of plastic hose as a level.

Although it is possible to start the transect on the upper shore and work down, it is much easier to use this piece of kit when working up the shore from low water mark. On Somerset shores, with a tidal range of over 10 m, cross staffs with $h = 1$ metre are easy to use (fig. 107). But in Shetland, where the tidal range is about 2 m, we had to use one with $h = 20$ cm. Using a cross-staff as small as this requires a degree of contortion that must have been amusing to watch – especially near low water mark.

If, for any reason, you want to run a transect across the shore at a constant height (around a boulder or in or out of a cave, for instance), a level could be made from a length of clear plastic hose, almost full of water (fig. 108).

Quadrats

Fig. 109. Using a quarter square metre framed quadrat to define a sampling area.

At each station along a transect, and in many other situations, it is necessary to examine the fauna and flora within a defined area (fig. 109). Such sampling areas are called quadrats, regardless of their actual shape. They may be of any convenient size, depending on the purpose of the study. For statistical purposes, it is better to collect data from a number of small quadrats than from one large one.

As snails exist as discrete individuals, we usually aim to record *density* (the number of individuals per unit area) rather than *cover* (the area of substrate occupied by the species). It is therefore helpful to know the area of your quadrat – a fact that probably explains the popularity of square quadrat frames! One of the sampling problems posed by rocky shores is the broken nature of the substrate, so that a 0.5 m × 0.5 m quadrat frame actually encloses an area of rock surface (much) greater than its theoretical quarter of a square metre (fig. 110). Moreover, as all small snails choose to occupy crevices in preference to flat areas, it may well be that the surface area within crevices is more important than that between crevices.

Fig. 110. Inappropriate use of a framed quadrat. The surface area of the rock within the frame is much greater than than the notional 0.25 m².

At one time, I favoured quadrat frames made of wood (figs 109 and 110, because they float), but latterly I have found those made of the white plastic pipe intended for overflow plumbing preferable as they are easier to make and carry.

Fig. 111. A quadrat frame designed for repeated sampling. Holes drilled in the wide border fitted over pegs set in the rock.

Fig. 112. Gore Point, in Somerset, where a small stream flows across the beach.

For some investigations, such as determining the rate of prey consumption by dog-whelks, it is necessary to relocate a quadrat frame in exactly the same place on different occasions. I have done this (fig. 111) by setting pegs into the rock and hanging my quadrat on them for each sampling session. The pegs were brass screws set with 'Araldite' into holes drilled into the rock with a masonry bit. The quadrat frame was made of galvanised zinc, wide enough to take the locating holes and for me to write in chalk the date and site name before taking a photograph of the sample area.

On broken shores composed of stable cobbles and boulders (such as fig. 112) it is almost meaningless to attempt to use a quadrat frame. Instead, I chose a surface upon which it was not too uncomfortable to kneel, and sampled all the snails that I could reach without moving my knees. Had I needed to record the area within my personal quadrat, I could have knelt on a floor, somewhere, and drawn the extent of my reach on a sheet of paper with a pencil. The result would be an interesting ellipse and the simplest way of determining the area would be to cut out the shape and weigh it, having also cut out and weighed a shape of known area drawn on the same type of paper.

Difficulties occur when recording densities up a transect, especially up a steep shore of small tidal range (which is usually the case in Shetland). You are surveying a wide strip of rock all at the same height; there is no flat surface on which to place a quadrat frame. Instead of trying to count all the individuals within a defined area, we used a variant (table 2) of the abundance scales devised by Crisp & Southward (1958) for their survey of the shores of the English Channel. Where a scale is labelled 'barnacles, small winkles' it means that it should be used for each species of barnacle (or winkle), in turn, and not that they should all be lumped together!

Where densities are so high that the individuals cannot be counted, as may be the case with tiny winkles living in and between the shells of barnacles (far exceeding the density of the barnacles) or where the snails have clustered in crevices and there is no way of finding

Table 2. Abundance scales for rocky shore surveys (modified after Crisp & Southward, 1958).

BARNACLES, SMALL WINKLES

E	500 or more/0.01 m^2
S	300–499/0.01 m^2
A	100–299/0.01 m^2
C	10–99/0.01 m^2
F	1–9/0.01 m^2
O	1–99/m^2
R	Less than 1/m^2
N	Absent

LIMPETS, LARGE WINKLES

E	20 or more/0.1 m^2
S	10–19/0.1 m^2
A	5–9/0.1 m^2
C	1–4/0.1 m^2
F	5–9/m^2
O	1–4/m^2
R	Less than 1/m^2
N	Absent

DOGWHELKS, TOPSHELLS

E	10 or more/0.1 m^2
S	5–9/0.1 m^2
A	1–4/0.1 m^2
C	5–9/m^2
F	1–4/m^2
O	Generally less than 1/m^2
R	Always less than 1/m^2
N	Absent

MUSSELS

E	More than 80% cover
S	50–79% cover
A	20–49% cover
C	5–19% cover
F	Small patches, less than 5% cover of rock surface
O	1–9 individuals/m^2
R	Less than 1/m^2
N	Absent

E = Extremely abundant,
S = Superabundant,
A = Abundant,
C = Common, F = Frequent,
O = Occasional, R = Rare and
N = Not seen.

out how many individuals are present, it can be more accurate to record *frequency* rather than to estimate *density*.

In this context, *frequency* may be defined as the chance of finding an individual within a given area. I made my quadrat (fig. 111) by drilling a series of holes through the border of 0.5 × 0.5 m frame and used fishing line to construct a lattice dividing the quarter of a square metre into 25 equal-sized squares. I recorded frequencies by counting the number of small squares in which each species was present.

This can be a very much quicker method of recording quantitative data, and invaluable when time is limiting. But it must always be remembered that (unlike counts) the results are dependent on the quadrat used. It is perfectly valid to compare frequencies obtained with the same quadrat in different places, but not frequencies obtained using quadrats of different sizes.

Species diversity

It is well known that estuarine shores harbour a rather monotonous fauna; animals may be abundant, but there are rather few species. By contrast, the clear-water shores of off-lying islands are much more exciting for biologists – almost every animal you find is different from the one before. Alas, human disturbance and pollution can significantly reduce the interest of sea shores (as well as of most other habitats).

If we are to identify the most exciting habitats so that they can be protected and cherished, and also those in need of remedial attention, we need some way of quantifying their biological value. One measure is *species richness* – a simple count of the number of species present. A more sophisticated measure, more useful for comparing sites, is *species diversity*.

An index of diversity incorporates a measure of the number of individuals of each species. Imagine two communities, each of one hundred snails, members of ten different species. A community made up of ten individuals from each of the ten species would be regarded as much more diverse than one in which one species was represented by 91 individuals and the others by just one each. There are several different indices of diversity (Wheater & Cook, 2003; Magurran, 2004). One that is often used is the Simpson Index, D, where $D = 1 - \sum p_i^2$, and p_i is the proportion of individuals represented by the ith species .

Snails, relatively easy to find and to identify, are very good subjects for studies of diversity, and an index of diversity based on the snail fauna can be a valuable aid in comparing two or more different shores, or even different parts of the same shore. Theoretically, we should expect species diversity to be lower in more stressful habitats and higher in more benign ones.

As mentioned earlier, there is an area of shore on Gore Point, in Somerset, where a small freshwater stream runs across the beach. The additional osmotic stress, thereby imposed on the shore fauna and flora, would be expected to lower species diversity. Calculation of the Simpson Index D at two different places (in the area of the stream and away from the stream) is illustrated by the imaginary (but not unrealistic) data in table 3, below. Begin by recording the number of individuals of each species found at each site when sampled in exactly the same manner. Then determine the proportion of the total number of individuals (p_i) represented by each species. These proportions are then squared (p_i^2) and summed ($\sum p_i^2$), and the result is subtracted from one ($1 - \sum p_i^2$).

Table 3. Imaginary data to illustrate calculation of the Simpson index of diversity, D.

Species	Site 1 in the area of the stream			Site 2 away from the stream		
	n	p_1	p_1^2	n	p_2	p_2^2
Patella vulgata	4	0.03	0.0009	8	0.05	0.0026
Littorina littorea	41	0.31	0.0979	25	0.16	0.0257
Littorina saxatilis	32	0.24	0.0597	45	0.29	0.0832
Littorina obtusata	0	0	0	4	0.03	0.0006
Gibbula umbilicalis	0	0	0	36	0.23	0.0532
Osilinus lineatus	54	0.41	0.1699	38	0.24	0.0593
Sum ($\sum$)	131	0.99	0.3285	156	1	0.2248
D	$D = 1 - \sum p_1^2 = \mathbf{0.67}$			$D = 1 - \sum p_2^2 = \mathbf{0.78}$		

Associations

As we saw on p. 39, flat winkles are thought to show food preferences for particular species of fucoid algae. The degree of association between a winkle and its putative food plant can be determined using Cole's coefficient of interspecific association. It is first necessary to construct a contingency table by examining a large number of small quadrats and recording the number that include one, the other or both of the two species.

		Species A present	Species A absent	
Species B	present	a	b	a + b
	absent	c	d	c + d
		a + c	b + d	a + b + c + d = n

Once figures have been obtained for a, b, c and d, it is a straightforward matter to calculate the degree of association between species A and species B. Cole's coefficient, C, is a number varying from +1 (complete positive association) through 0 (no association) to –1 (complete negative association). It, and its standard deviation, can be calculated by using one of the following formulae.

When $ad \geq bc$:

$$C_{AB} = \frac{ad - bc}{(a + b)(b + d)} \pm \sqrt{\frac{(a + c)(c + d)}{n(a + b)(b + d)}}$$

When $bc > ad$ and $d \geq a$:

$$C_{AB} = \frac{ad - bc}{(a + b)(a + c)} \pm \sqrt{\frac{(b + d)(c + d)}{n(a + b)(a + c)}}$$

When $bc > ad$ and $a > d$:

$$C_{AB} = \frac{ad - bc}{(b + d)(c + d)} \pm \sqrt{\frac{(a + b)(a + d)}{n(b + d)(c + d)}}$$

Salinity tolerance experiments

In the aftermath of the Torrey Canyon oil spill in 1967, when the cleanup operation caused far greater devastation to marine life than did the untreated oil, the Oil Pollution Research Unit was tasked with comparing the toxicity of the available detergents. They rapidly discovered that the level of dilution of the chemical was of critical importance and so developed a technique for measuring the effect of salinity on animal behaviour. At Nettlecombe Court Field Centre, we adapted the experiment to investigate the distribution of organisms in and around the Gore Point stream.

The experimental animals spent the first hour of their trial in 100 ml of water at one of nine different salinities - 0% sea water, 25%, 50%, 75%, 100%, 125%, 150%, 175% and 200%. Water collected from the sea at that site on that day was taken to be 100% sea water. Nettlecombe spring water was taken to be 0% sea water and these two sources were mixed appropriately to produce the intermediate solutions. Normal sea water is usually taken to be 35 gl^{-1} NaCl. Double strength sea water (70 gl^{-1} NaCl) can be manufactured by boiling sea water to half volume, but the resulting water must then be cooled and aerated. It is quicker and easier to make it by adding an extra 35 g of common salt to a litre of sea water. Concentrations ranging between 100% and 200% sea water were obtained by mixing those two as appropriate.

The relevance of the range between 0% and 100% to the stream-across-the-beach scenario should be obvious, but that of the range between 100% and 200% may be less so. As soon as a rockpool is exposed by the ebb tide on a warm windy day, water will begin to evaporate and salinities well in excess of 100% sea water must be commonplace; some small shallow pools evaporate to dryness and salt crystals are left in small hollows (fig. 113).

Fortyfive specimens of each animal species are required, five for each solution. (I used to advise my students to collect 50; it was surprising how many empty shells occupied by hermit crabs came back to the lab!) The specimens should, as far as possible, be all of the same size and have been treated in a uniform manner during the (short) interval between collection and the experiment. At the start, the experimental animals were placed in their designated bowls, upside-down. Taking great care to avoid vibration, excessive shading or fluctuating illumination of the bowls, the activity of each

Fig. 113. Salt crystals on the rock where the water in a small pool has evaporated on a warm windy day.

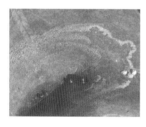

Table 4. Activity scale used in salinity tolerance experiments.

Operculum shut	0
Operculum open	1
Foot extended	2
Animal turned over	3
Animal crawling	4
Animal crawled out	5

Fig. 114. A pseudo three dimensional plot of the experimental phase of the salinity-tolerance behaviour experiment. The activity values (table 4) of the five snails in each pot were assessed every five minutes and summed to provide the data points plotted on the graph. (I find it easiest to plot such graphs 'backwards', starting with the final set of data (shown in black), then adding the points derived 5 minutes earlier etc etc, omitting the ones that would be invisible 'over the top of the hill.')

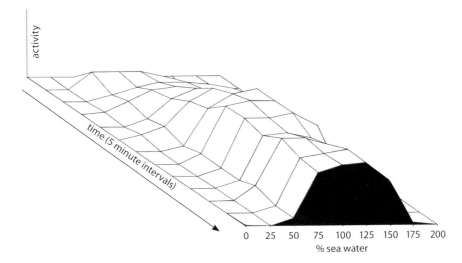

individual was recorded on a five-point activity scale (table 4) every five minutes for one hour.

At the end of that time, the animals were removed from the bowls but kept in their nine groups. The experimental solutions were discarded, the bowls were wiped out, and all were refilled with 100% sea water. The animals were replaced in the bowls and the observations repeated for a second hour. During this 'recovery phase' we recorded the effect on the animals' activity in normal sea water caused by the hour spent in the experimental solution.

Such data may be displayed by various forms of pseudo three dimensional graph to show the emergence of the pattern with time but an indication of the overall view may be obtained by summing the totals for each bowl.

All species that I have yet exposed to this experiment showed a peak of activity at or near 100% sea water, individuals in 0% and 200% usually remaining inactive (as in fig. 114.). During the recovery phase, those that were

Fig. 115. A pseudo three dimensional plot of the recovery phase of the salinity-tolerance behaviour experiment. Initially, there are three peaks of activity, amongst the snails that had been totally inactive in 0% and 200% sea water plus those that had been in 100–125% sea water. But, as the other animals recovered, there was much less difference at the end of the hour.

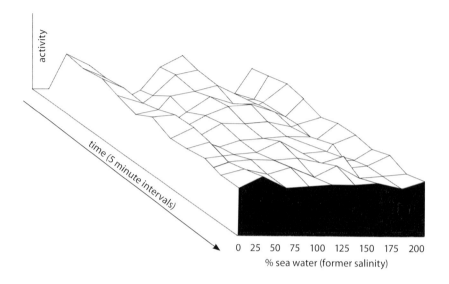

activity

time (5 minute intervals)

0   25   50   75   100   125   150   175   200

% sea water (former salinity)

previously inactive now behaved like the animals originally placed in 100% sea water whereas those that had been partially active during the first hour took a little while to recover from the experience (fig. 115). By the end of the second hour, all were behaving normally.

Often, the controls (the individuals in 100% sea water throughout) were less active during the second hour than during the first. But then, the excitement of crawling around a small glass bowl is limited, even for a topshell!

Measuring shells

When measuring limpets, do not remove them from the rock, as injury is almost inevitable. Limpet shells are most easily measured with a pair of dividers and a ruler for, except on the most even of surfaces, callipers are awkward to use. Shell height (h in fig. 116) is difficult to measure directly, but a simple ratio of frontal length (f) to overall length (l) may be calculated easily; higher values indicate more conical shells. The relationship between this ratio and h may be established from measurements taken on empty shells.

Fig. 116. Direct measurement of shell height (h) of limpets in their homes is often difficult. It is much easier to measure frontal slope (f – the shell apex is over the head) and length (l).

Fig. 117. Length (*L*) and aperture length (*Ap*); the two least equivocal measurements to take on a dog-whelk shell.

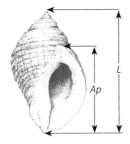

For helically coiled shells (for example, *Nucella lapillus*, fig. 117) I have found the two most objective measurements to take are length (*L*) and aperture length (*Ap*). Length is the maximum dimension across the shell involving the apex whilst aperture length is taken from the point where the suture ends (the upper end of the outside of the shell lip) to the opposite end of the aperture. It is much more difficult to define 'breadth' unequivocally – even on this two dimensional drawing. When handling a real shell it is easy to obtain a wide range of 'breadths' from a single specimen.

Marking shells

The well-known method of estimating population size through a process of marking, release and recapture requires marking the shells of living snails in a manner that will not harm the animal, affect its behaviour or render it more or less likely to suffer from predation. It goes without saying that the person who marked the shell must be able to recognise the marks at a later date but only after recapture – it must not be possible to bias the result by selecting marked individuals preferentially.

I once asked a group of students to mark a large number of winkles by dabbing a blob of paint on the upper surface of the shell. Returning to the shore the following day we met a small boy carrying a bucket. "Have you seen these funny snails?" So the idea is to mark the underside of the shell with a *small* dab of paint, which must not run down into the aperture and which must be allowed to dry before the animal emerges from its shell or the tide comes in.

The fast-drying gloss paints used by model makers are ideal for marking snails, and are available in very small tins. For the paint to bond with the shell surface, this must be dry and free from mucus.

The whole process of marking takes quite a long time and is most likely to be successful with an upper-shore species that is accustomed to dry conditions and is content to stay put when left upside down on the rocks for an hour or more whilst the paint is drying.

Shells marked in this way are recognisable for days, weeks or months – a few for even longer – but if it is

Fig. 118. The shell of a common topshell marked permanently, the previous summer, with a notch sawn into the shell lip.

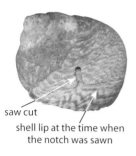

saw cut

shell lip at the time when the notch was sawn

necessary to be able to recognise an individual snail in later years a more permanent method is required. It is possible, though laborious, to cut a notch into the shell lip with a hacksaw. I have done this with dogwhelks when trying to establish growth rates and longevity, as has Dr M. A. Kendall with common topshells (fig. 118). The snail quickly fills in the notch, which none the less remains visible as it moves up the outer surface of the body whorl.

Estimating population size

The estimation of population size using the Lincoln Index following the marking, release and recapture of a sample is much favoured by the writers of text books (for example, Chalmers & Parker, 1989) but in practice there are few animal species to which it can be applied; the common topshell may be an exception.

A sample of the animals (S_1) is collected and the individuals marked as discussed above. The marked snails are released back into their habitat and allowed time to mingle at random with other members of the population. A second sample (S_2) is then taken, at random, and divided into marked (R) and unmarked individuals. The estimate of population size (N) is based on the presumption that the ratio of recaptures from the first sample to the whole of the second sample ($R:S_2$) is the same as the ratio of all the marked individuals to the total population ($S_1:N$). So,

$$N = (S_1 \times S_2) \div R$$

The 95% confidence limits of the estimate (within which lies the true value of N) are given by

$$N \pm 2N \sqrt{(1 \div R) - (1 \div S_2)}$$

It is an elegant concept and the arithmetic presents no problem. It assumes, however, that individuals can be sampled, handled and marked without harm and without affecting their behaviour; that the habitat is not damaged by the sampling process; that released individuals mix at random within their population; that there are no births or deaths, immigrations to or emigrations from the study area, during that period; and that marked individuals are neither more nor less likely (than unmarked ones) to be recaptured when the second sample is taken.

The technique is obviously inappropriate for territorial animals, such as limpets, or for those forever on the move, such as shore crabs.

Most attempts to apply this technique using terrestrial invertebrates are seriously flawed because the assumptions cannot be met. Not many people can hold insects, and mark them, without damaging them and without smearing the paint across the body. The only time I tried the technique with land snails, we destroyed the habitat (a nettlebed) collecting the first sample and could not persuade the snails to remain inactive until the paint had dried. (A schoolmaster friend of mine achieved a certain notoriety by using the technique as a class exercise to investigate an adder population! The snakes were slid, headfirst, into a clear plastic tube into which holes had been drilled to facilitate adding spots of paint.)

But the real problem is the requirement for marked individuals to mingle at random within the unmarked population. It is hard to imagine that adult males and females are ever distributed at random within most populations.

The common topshell, *Osilinus lineatus,* is a suitable subject for this technique because:

• it is easy to collect the initial sample without disturbing the habitat;

• they are easy to mark with a small dab of paint on the underside of the shell so that the mark is not visible until the animal is picked up and turned over;

• they can be left in full sun while the paint dries;

• they are active mobile animals which move about readily;

• males and females do not copulate (the male does not have a penis); both sexes release their gametes into the sea and fertilisation occurs in the water. No courtship behaviour has been described, so individuals of this species really may mingle at random.

I contend that it is not possible to collect a sample of snails 'at random' when sampling involves searching as distinct from merely recording. The best we can do is to collect 'without conscious bias' and I did this by marking out an area and collecting all the individuals that could be found within it.

On 2 June 1996, a 20 m × 20 m square was marked on the shore of Gore Point and twenty-five people collected, marked and released 2,274 common topshells within that square. (About 800 others were released unmarked because the tide would have covered them before the paint had had time to dry.) The following day, a second collection was made. The 2,493 common topshells collected in the marked area included 1,230 marked individuals. On the basis of these numbers, we estimated the population to be 4,609 or 12/m², which seemed to be of the right order of magnitude. However, at least 25 marked animals had already left the marked area. The second collection was released back into the marked area.

A week after the original release of marked animals, a third collection was made. This time, the 2,539 common topshells collected included only 579 recaptures; on the basis of these numbers, the estimated population size had risen to 9,972 or 25/m². That estimate must have been distorted by the rate of emigration of marked animals from and immigration of unmarked ones into the square, as the size of the collections, each made by similar numbers of people at the same marked site in the same amount of time, shows that there were actually about the same number present on all three occasions.

This example highlights the problem of deciding when to recapture; too soon, and the marked individuals will not have had time to mingle with unmarked members of the population; too late, and immigration and emigration may seriously distort the result. For this species, in summer, it seems as though 24 hours (allowing two tides to cover the site) may be the optimal interval between collections.

I imagine that the technique would be equally applicable to purple topshells, *Gibbula umbilicalis*, but perhaps not to the larger winkles. The procedure would be easy enough to perform but I doubt whether they would ever mingle at random. Male rough winkles seem to be forever searching for receptive females.

Dog-whelks move about very little (I have recaptured marked individuals within 30 cm of their release point a year after their release) and so it would be much more appropriate simply to count them.

Fig. 119. A running-mean plot to determine the sample size required to measure the mean shell shape of a dog-whelk enclave. From this, and similar plots, I determined that samples of 30 individuals would be as useful as samples of 100.

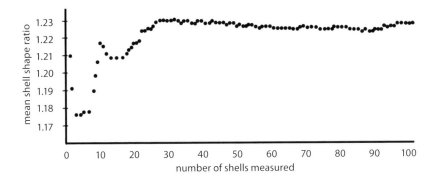

Deciding on the optimal sample size

When describing the measurement of dog-whelk shell shape (p. 33), I mentioned that I considered 30 adult individuals, collected without conscious bias, to be a large enough sample from which to determine the mean. I ought to explain why.

I started off by taking samples of 100 shells; a nice round number and one that I was sure would be sufficient. It was, however, somewhat laborious on shores where the species was sparse. So I plotted graphs of the kind reproduced as fig. 119. This is a running mean of the value of the L/Ap ratio (fig. 117) against the number of snails measured. It will be seen that the mean did not change significantly after 30 individuals had been measured. This graph is of an exposed shore population, from Long Nose on Skokholm, but similar results were obtained from sheltered shores as well.

Pellet analysis

Fig. 120. A gull pellet.

Modern birds do not have teeth, or powerful crushing jaws, and so they have to swallow rocky shore snails whole. Potential food material is ground up in the ventriculus (gizzard) or stomach. At the end of this process, the edible component, now in semi-liquid form, passes on through a sphincter into the intestine whilst the hard, inedible, residue is regurgitated as a pellet (*e.g.* fig. 120).

This one appears to be composed of crab skeletal remains, but others often contain shells of rough winkles. Those of eiders, consisting primarily of crushed mussel shells, sometimes contain dog-whelk shells.

Examination of pellets is rarely a smelly or unpleasant task, very different from the examination of faeces, and can be highly informative regarding not only the species being selected as food but also the preferred size range.

Gulls may deposit their pellets anywhere (this one was found on a sandy beach) but most individuals have favoured vantage points from which they watch the world go by and where they regurgitate most of their pellets.

10 References and further reading

Some of the books and journals listed here will be unavailable in local and school libraries. It is possible to make arrangements to see or borrow such works in several ways: by seeking permission to visit the library of a local university or natural history society, or by asking your local public library to borrow the work (or a photocopy of it) for you via the British Library, Document Supply Centre. This may take several weeks, and it is important to present your librarian with a reference that is correct in every detail. References are acceptable in the form given here, namely the author's name and date of publication, followed by (for a book) the title and publisher or (for a journal article) the title of the article, the journal title, the volume number, and the first and last pages of the article.

Increasingly, the publishers of scientific journals are making their archive material freely available on the internet. Most of the papers that appeared in *Field Studies*, for example, are available via the Field Studies Council website (www.field-studies-council.org); simply click on 'Publishing.' It is to be hoped that all librarians will be in a position to advise but, if in doubt, I suggest searching for the journal title on the internet.

There are two scientific societies devoted to molluscs. The Conchological Society of London was originally an amateur body primarily interested in shells whilst the Malacological Society of London was the professional body more interested in soft parts of the animals – but these distinctions have become blurred. The Conchological Society publishes the *Journal of Conchology* and *Mollusc World*. From 1976, *The Proceedings of the Malacological Society of London* have been issued as *The Journal of Molluscan Studies*.

Admiralty (annually). *The Admiralty Tide Tables volume 1 European Waters*. HMSO.

Atkinson, W. D. & Warwick, T. (1983). The role of selection in the colour polymorphism of *Littorina rudis* Maton and *Littorina arcana* Hannaford Ellis (Prosobranchia: Littorinidae). *Biological Journal of the Linnean Society*, **20**: 137–151.

Baker, J. M. & Crothers, J. H. (1987). *Intertidal rock*. Chapter 8 of Baker, J. M. & Wolff, W. J. (eds) *Biological surveys of estuaries and coasts*. Cambridge University Press, Cambridge.

Ballantine, W. J. (1961). A biologically-defined exposure scale for the comparative description of rocky shores. *Field Studies*, **1**(3): 1–19.

Bassindale, R. (1941). Studies on the biology of the Bristol Channel IV. The invertebrate fauna of the southern shores of the Bristol Channel and Severn Estuary. *Proceedings of the Bristol Naturalists' Society*, **9**(1940): 143–201.

Bassindale, R. (1943). Studies on the biology of the Bristol Channel XIII. The intertidal fauna of Porlock Bay. *Proceedings of the Bristol Naturalists' Society*, **9**(1942): 386–399.

Berry, R. J. & Crothers, J. H. (1968). Stabilizing selection in the dog-whelk, *Nucella lapillus*. *Journal of Zoology*, **155**: 5–17.

Berry, R. J. & Crothers, J. H. (1970). Genotypic stability and physiological tolerance in the dog-whelk, *Nucella lapillus*. *Journal of Zoology*, **162**: 293–301.

Berry, R. J. & Crothers, J. H. (1974). Visible variation in the dog-whelk, *Nucella lapillus*. *Journal of Zoology*, **174**: 123–148.

Boaventura, D., da Fonseca, L. C. & Hawkins, S. J. (2003). Size matters: competition within populations of the limpet *Patella depressa*. *Journal of Animal Ecology*, **72**: 435–446.

Bryan, G. W., Gibbs, P. E., Hummerstone, L. G. & Burt, G. R. (1986). The decline of the gastropod *Nucella lapillus* around south-west England: evidence for the effect of tributyltin from anti-fouling paints. *Journal of the Marine Biological Association of the U.K.*, **66**: 525–544.

Byers, B. A. (1990). Shell colour polymorphism associated with substrate colour in the intertidal snail *Littorina saxatilis* Olivi. *Biological Journal of the Linnean Society*, **40**: 3–10.

Carefoot, T. (1977). *Pacific seashores. A guide to intertidal ecology*. J. J. Douglas, Vancouver.

Chalmers, N. & Parker, P. (1989). *The OU project guide: fieldwork and statistics for ecological projects (second edition)*. Field Studies Council, Shrewsbury. No 9 in a series of occasional publications.

Cowell, E. B. & Crothers, J. H. (1970). On the occurrence of multiple rows of 'teeth' in the shell of the dog-whelk, *Nucella lapillus*. *Journal of the Marine Biological Association, U.K.*, **50**: 1101–1111.

Crisp, D. J. (ed.) (1964). The effects of the severe winter 1962–1963 on marine life in Britain. *Journal of Animal Ecology*, **33**: 165–210.

Crisp, D. J., & Knight-Jones, E. W. (1955). Discontinuities in the distribution of shore animals in North Wales. *Bardsey Observatory Report*, 1954: 29–34.

Crisp, D. J., & Southward, A. J. (1958). The distribution of intertidal organisms along the coasts of the English Channel. *Journal of the Marine Biological Association, U.K.*, **37**: 157–208.

Crothers, J. H. (1967). The biology of the shore crab, *Carcinus maenas* (L.). 1. The background – anatomy, growth and life history. *Field Studies*, **2**: 407–434.

Crothers, J. H. (1968). The biology of the shore crab, *Carcinus maenas* (L.). 2. The life of the adult crab. *Field Studies*, **2**: 579–614.

Crothers, J. H. (1971). Further observations on the occurrence of "teeth" in the dog-whelk, *Nucella lapillus* (L.). *Journal of the Marine Biological Association, U.K.*, **51**: 623–629.

Crothers, J. H. (1973). On variation in *Nucella lapillus* (L.): shell shape in populations from Pembrokeshire, South Wales. *Proceedings of the Malacological Society of London*, **40**: 319–327.

Crothers, J. H. (1974). On variation in *Nucella lapillus* (L.): shell shape in populations from the Bristol Channel. *Proceedings of the Malacological Society of London*, **41**: 157–170.

Crothers, J. H. (1975). On variation in *Nucella lapillus* (L.): shell shape in populations from the south coast of England. *Proceedings of the Malacological Society of London*, **41**: 489–498.

Crothers, J. H. (1976). On the distribution of some common animals and plants on rocky shores of West Somerset. *Field Studies*, **4**: 369–389.

Crothers, J. H. (1981). On the graphical presentation of quantitative data. *Field Studies*, **5**: 487–511.

Crothers, J. H. (1983). Field experiments on the effects of crude oil and dispersant on common animals and plants of rocky sea shores. *Marine Environmental Research*, **8**: 215–239.

Crothers, J. H. (1985a). Two different patterns of shell-shape variation in the dog-whelk, *Nucella lapillus* (L.). *Biological Journal of the Linnean Society*, **25**: 339–353.

Crothers, J. H. (1985b) Dog-whelks : an introduction to the biology of *Nucella lapillus* (L.). *Field Studies*, **6**: 291–360.

Crothers, J. H. (1992). Shell size and shape variation in edible winkles, *Littorina littorea* (L.), from West Somerset. *Proceedings of the Third International Symposium on Littorinid Biology*, 91–97.

Crothers, J. H. (1998). A hot summer, two cold winters and the geographical limit of *Trochocochlea lineata* in Somerset. *Hydrobiologia*, **378**: 133–141.

Crothers, J. H. (2001). Common topshells: an introduction to the biology of *Osilinus lineatus* with notes on other species in the genus. *Field Studies*, **10**: 115–160.

Crothers, J. H. (2003a). Further observations on a population of dog-whelks, *Nucella lapillus* (Gastropoda) recolonizing a site following amelioration of tributyltin (TBT) pollution. *Journal of the Marine Biological Association, UK*, **83**: 1023–1027.

Crothers, J. H. (2003b). Rocky shore snails as material for projects (with a key for their identification). *Field Studies*, **10**: 601–634.

Crump, R. G., Williams, A. D. & Crothers, J. H. (2003). West Angle Bay: a case study. The fate of limpets. *Field Studies*, **10**: 579–599.

Dalby, D. H., Cowell, E. B., Syratt, W. B. & Crothers, J. H. (1978).

An exposure scale for marine shores in Western Norway. *Journal of the Marine Biological Association, U.K.*, **58**: 975–996.

Desai, B. N. (1966). The biology of *Monodonta lineata* (da Costa). *Proceedings of the Malacological Society of London*, **37**: 1–17.

Etter, R. J. (1988). Physiological stress and colour polymorphism in the intertidal snail *Nucella lapillus*. *Evolution*, **42**: 660–680.

Feare, C. J. (1970). Aspects of the ecology of an exposed shore population of dogwhelks *Nucella lapillus*. *Oecologia (Berlin)*, **5**: 1–18.

Fish, J. D. (1972). The breeding cycle and growth of open coast and estuarine populations of *Littorina littorea*. *Journal of the Marine Biological Association, UK*, **52**: 1011–1109.

Fretter, V. & Graham, A. (1962; 1994). *British prosobranch molluscs*. Ray Society, London.

Fretter, V. & Graham, A. (1976). The prosobranch molluscs of Britain and Denmark Part 1 — Pleurotomariacea, Fissurellacea and Patellacea. *Journal of Molluscan Studies*, supplement 1.

Fretter, V. & Graham, A. (1977). The prosobranch molluscs of Britain and Denmark Part 2 — Trochacea. *Journal of Molluscan Studies*, supplement 3.

Fretter, V. & Graham, A. (1980). The prosobranch molluscs of Britain and Denmark Part 5 — Marine Littorinacea. *Journal of Molluscan Studies*, supplement 5.

Fretter, V. & Graham, A. (1985). The prosobranch molluscs of Britain and Denmark Part 8 — Neogastropoda. *Journal of Molluscan Studies*, supplement 15.

Fretter, V. & Pilkington, M. C. (1970). Prosobranchia. Veliger larvae of Taenioglossa and Stenoglossa. *Fiches d'identification du zooplankton*, sheets 129–132.

Fretter, V. & Shale, D. (1973). Seasonal changes in population density and vertical distribution of prosobranch veligers in offshore plankton at Plymouth. *Journal of the Marine Biological Association, UK*, **53**: 471–492.

Garwood, P. R., & Kendall, M. A. (1985). The reproductive cycles of *Monodonta lineata* and *Gibbula umbilicalis* on the coast of Mid-Wales. *Journal of the Marine Biological Association, UK*, **65**: 993–1008.

Gibbs, P. E. (1993). Phenotypic changes in the progeny of *Nucella lapillus* (Gastropoda) transplanted from an exposed shore to sheltered inlets. *Journal of Molluscan Studies*, **59**: 187–194.

Graham, A. (1988). *Molluscs: prosobranch and pyramidellid gastropods*. Synopsis of the British fauna No. 2 (second edition) published for the Linnean Society of London and the Estuarine and Brackish-water Sciences Association by E. J. Brill/ Dr W. Backhuys, Leiden.

Goodwin, B. J. & Fish, J. D. (1977). Inter- and intra-specific variation in *Littorina obtusata* and *L. mariae* (Gastropoda; Prosobranchia). *Journal of molluscan studies*, **43**: 241–254.

Hannaford Ellis, C. J. (1978). *Littorina arcana* sp. nov.: a new species of winkle (Gastropoda: Prosobranchia: Littorinidae). *Journal of Conchology*, **29**: 304.

Hannaford Ellis, C. J. (1985). The breeding migration of *Littorina arcana* Hannaford Ellis, 1978 (Prosobranchia: Littorinidae). *Zoological Journal of the Linnean Society*, **84**: 91–96.

Hawkins, S. J. & Hiscock, K. (1983). Anomalies in the abundance of common eulittoral gastropods with planktonic larvae on Lundy, Bristol Channel. *Journal of Molluscan Studies*, **49**: 86–88.

Hawthorne, J. B. (1965). The eastern limit of distribution of *Monodonta lineata* (da Costa) in the English Channel. *Journal of Conchology*, **25**: 384–352.

Hayward, P. J. & Ryland, J. S. (1990). *The marine fauna of the British Isles and north-west Europe*. Two volumes. The Clarendon Press, Oxford.

Hayward, P. J. & Ryland, J. S. (1995). *Handbook of the marine fauna of north-west Europe*. Oxford University Press, Oxford.

Heller, J. (1975a). The taxonomy of some British *Littorina* species with notes on their reproduction. *Zoological Journal of the Linnean Society*, **56**: 131–151.

Heller, J. (1975b). Visual selection of shell colour in two littoral prosobranchs. *Zoological Journal of the Linnean Society*, **56**: 153–170.

Hoagland, KE. (1977). A gastropod color polymorphism: one adaptive strategy of phenotypic variation. *Biological Bulletin*, **152**: 360–372.

Janson, K. & Ward, R. D. (1984). Microgeographic variation in allozyme and shell characters in *Littorina saxatilis* Olivi (Prosobranchia: Littorinidae). *Biological Journal of the Linnean Society*, **22**: 289–307.

Jones, N. S. (1948). Observations and experiments on the biology of *Patella vulgata* at Port St Mary, Isle of Man. *Proceedings and Transactions of the Liverpool Biological Society*, **56**: 60–77.

Kendall, M. A. (1987). The age and size structure of some northern populations of the trochid gastropod *Monodonta lineata*. *Journal of Molluscan Studies*, **53**: 213–222.

Kendall, M. A., Williamson, P., & Garwood, P. R. (1987). Annual variation in recruitment and population structure of *Monodonta lineata* and *Gibbula umbilicalis* populations at Aberaeron, mid Wales. *Estuarine and Coastal Shelf Science*, **24**: 499–511.

Lancaster, I. (1988). *Pagurus bernhardus* (L.) – an introduction to the natural history of hermit crabs. *Field Studies*, **7**: 189–238.

Little, A. E., Dicks, B. & Crothers, J. H. (1986). Studies of barnacles, limpets and topshells in Milford Haven. *Field Studies*, **6**: 459–492.

Little, C. (1989). Factors governing patterns of foraging activity in littoral marine herbivorous molluscs. *Journal of Molluscan Studies*, **55**: 273–284.

Little, C. & Kitching, J. A. (1996). *The biology of rocky shores*. Oxford University Press, Oxford.

Little, C., Morritt, D., Paterson, D. M., Stirling, P. & Williams, G. A. (1990). Preliminary observations on factors affecting foraging activity in the limpet *Patella vulgata*. *Journal of the Marine Biological Association of the UK*, **70**: 181–195.

Little, C., Partridge, J. C. & Teagle, E. (1991). Foraging activities in normal and abnormal tidal regimes. *Journal of the Marine Biological Association of the UK*, **71**: 537–554.

Magurran, A. (2004). *Measuring biological diversity*. Blackwell Science Ltd, Oxford.

Moyse, J. & Nelson-Smith, A. (1964). Effects of the severe cold of 1962–63 upon shore animals in South Wales. *Journal of Animal Ecology*, **33**: 1–31.

Orton, J. H. (1929). Observations on *Patella vulgata*. part 3. Habitat and habits. *Journal of the Marine Biological Association UK*, **16**: 277–288.

Pettitt, C. W. (1975). A review of the predators of *Littorina*, especially those of *L. saxatilis* (Olivi) (Gastropoda: Prosobranchia). *Journal of Conchology*, **28**: 343–357.

Preece, R. C. (1993). *Monodonta lineata* (da Costa) from archaeological sites in Dorset and the Isle of Wight. *Journal of Conchology*, **34**: 339–341.

Raffaelli, D. (1979). Colour polymorphism in the intertidal snail *Littorina rudis* Maton. *Zoological Journal of the Linnean Society*, **67**: 65–73.

Rainbow, P. S. (1984). An introduction to the biology of British barnacles. *Field Studies*, **6**: 1–51.

Reid, D. G. (1996). *Systematics and evolution of Littorina*. The Ray Society, London.

Reimchen, T. E. (1979). Substratum heterogeneity, crypsis, and colour polymorphism in an intertidal snail (*Littorina mariae*). *Canadian Journal of Zoology*, **57**: 1070–1085.

Rex, M. A., Stuart, C. T., Etter, R. J. & McClain, C. R. (2010). Biogeography of the deep-sea gastropod *Oocorys sulculata* Fischer 1884. *Journal of conchology*, **40**: 287–290.

Scheltema, R. S. (1971). Larval dispersal as a means of genetic exchange between geographically separated populations of shallow-water benthic marine gastropods. *Biological Bulletin, Woods Hole*, **140**: 284–322.

Smith, D. A. S. (1976). Disruptive selection and morph-ratio clines in the polymorphic snail *Littorina obtusata* (L.) (Gastropoda: Prosobranchia). *Journal of Molluscan Studies*, **42**: 114–135.

Southward, A. J. (2007). *Barnacles*. Synopsis of the British fauna No. 57 published for the Linnean Society of London and the Estuarine and Coastal Sciences Association by Field Studies Council, Shrewsbury.

Steneck, R. S. & Watling, L. (1982). Feeding capabilities and limitations of herbivorous molluscs: a functional group approach. *Marine Biology*, **68**: 299–319.

Torunski, H. (1979). Biological erosion and its significance for the morphogenesis of limestone coasts and for nearshore sedimentation (Northern Adriatic). *Senckenbergiana maritima*, **11**: 193–265.

Underwood, A. J. (1973). Studies on zonation of intertidal prosobranch molluscs in the Plymouth region. *Journal of Animal Ecology*, **42**: 353–372.

Waite, M. E., Waldock, M. J., Thain, J. E., Smith, D. J. & Milton, S. M. (1991). Reductions in TBT concentrations in UK estuaries following legislation in 1986 and 1987. *Marine Environmental Research*, **32**: 89–111.

Wheater, C. P. & Cook, P. A. (2003). *Studying invertebrates*. Naturalists' Handbooks 28. The Richmond Publishing Co. Ltd, Slough.

Williams, E. E. (1965). The growth and distribution of *Monodonta lineata* (da Costa) on a rocky shore in Wales. *Field Studies*, **2**: 189–198.

Williams, G. A. (1990). The comparative ecology of the flat periwinkles, *Littorina obtusata* (L.) and *L. mariae* Sacchi et Rastelli. *Field Studies*, **7**: 469–482.

Williamson, P. & Kendall, M. A. (1981). Population age structure and growth of the trochid *Monodonta lineata* determined from shell rings. *Journal of the Marine Biological Association, UK*, **61**: 1011–1026.

Wilson, C. M., Crothers, J. H. & Oldham, J. H. (1983). Realized niche: the effects of a small stream on seashore distribution patterns. *Journal of Biological Education*, **17**: 51–58.

11 Index

12 Pictorial key

One finding of the AIDGAP project (Aids to Identification in Difficult Groups of Animals and Plants) was that whilst people who had been taught to use dichotomous keys (for example, pages 48–57 of this book) as students usually preferred that format, people without such training often preferred a more pictorial arrangement. As some users of this book may be of the same opinion, essentially the same information has been re-arranged on the following pages.

It will be necessary to refer to fig. K1 (on page 48) for definitions of the technical terms used to describe shell ornament and to figs K2 and K3 for illustrations of opercula.

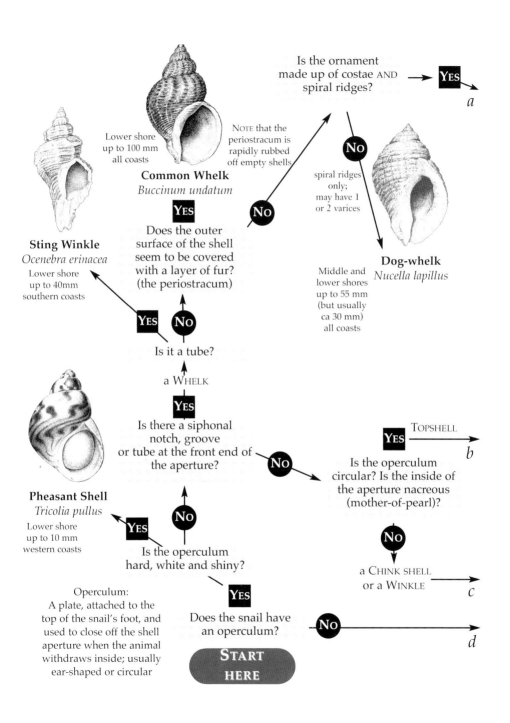

Is the ornament made up of costae AND spiral ridges? → **YES** → *a*

NO

spiral ridges only; may have 1 or 2 varices

NOTE that the periostracum is rapidly rubbed off empty shells

Lower shore up to 100 mm all coasts

Common Whelk
Buccinum undatum

YES

NO

Dog-whelk
Nucella lapillus

Middle and lower shores up to 55 mm (but usually ca 30 mm) all coasts

Does the outer surface of the shell seem to be covered with a layer of fur? (the periostracum)

Sting Winkle
Ocenebra erinacea
Lower shore up to 40mm southern coasts

YES   **NO**

Is it a tube?

a WHELK

YES

Is there a siphonal notch, groove or tube at the front end of the aperture?   **NO**

YES TOPSHELL → *b*

Is the operculum circular? Is the inside of the aperture nacreous (mother-of-pearl)?

NO

a CHINK SHELL or a WINKLE → *c*

Pheasant Shell
Tricolia pullus
Lower shore up to 10 mm western coasts

NO

YES

Is the operculum hard, white and shiny?

Operculum: A plate, attached to the top of the snail's foot, and used to close off the shell aperture when the animal withdraws inside; usually ear-shaped or circular

YES

Does the snail have an operculum?   **NO** → *d*

START HERE

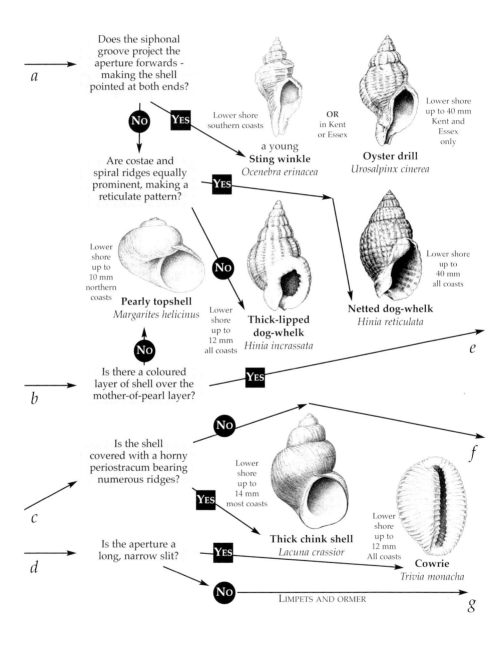

a

Does the siphonal groove project the aperture forwards - making the shell pointed at both ends?

No **Yes**

a young
Sting winkle
Ocenebra erinacea

Lower shore southern coasts

OR
in Kent or Essex

Oyster drill
Urosalpinx cinerea

Lower shore up to 40 mm Kent and Essex only

Are costae and spiral ridges equally prominent, making a reticulate pattern?

Yes

No

Pearly topshell
Margarites helicinus

Lower shore up to 10 mm northern coasts

Thick-lipped dog-whelk
Hinia incrassata

Lower shore up to 12 mm all coasts

Netted dog-whelk
Hinia reticulata

Lower shore up to 40 mm all coasts

e

No

b

Is there a coloured layer of shell over the mother-of-pearl layer?

Yes

No

f

c

Is the shell covered with a horny periostracum bearing numerous ridges?

Yes

Thick chink shell
Lacuna crassior

Lower shore up to 14 mm most coasts

d

Is the aperture a long, narrow slit?

Yes

Cowrie
Trivia monacha

Lower shore up to 12 mm All coasts

No

LIMPETS AND ORMER

g

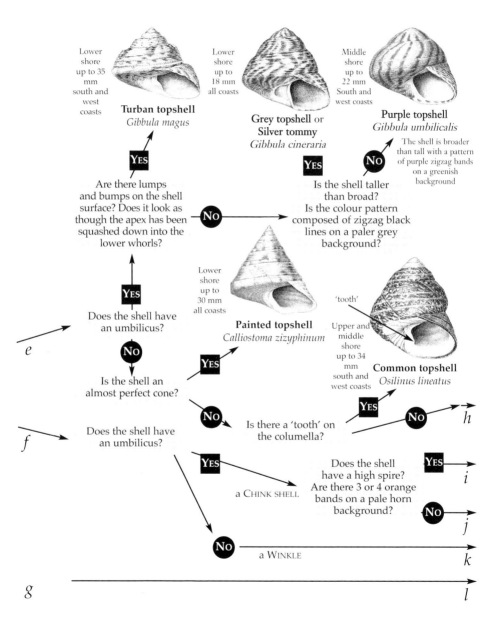

Lower shore up to 35 mm south and west coasts

Turban topshell
Gibbula magus

Lower shore up to 18 mm all coasts

Grey topshell or **Silver tommy**
Gibbula cineraria

Middle shore up to 22 mm South and west coasts

Purple topshell
Gibbula umbilicalis

The shell is broader than tall with a pattern of purple zigzag bands on a greenish background

Yes

Are there lumps and bumps on the shell surface? Does it look as though the apex has been squashed down into the lower whorls?

No →

Yes

Is the shell taller than broad? Is the colour pattern composed of zigzag black lines on a paler grey background?

No

Yes

Does the shell have an umbilicus?

No

Is the shell an almost perfect cone?

Lower shore up to 30 mm all coasts

Painted topshell
Calliostoma zizyphinum

Yes

'tooth'

Upper and middle shore up to 34 mm south and west coasts

Common topshell
Osilinus lineatus

Yes **No** →

h

e →

f →

Does the shell have an umbilicus?

No →

Is there a 'tooth' on the columella?

Yes →

a CHINK SHELL

Does the shell have a high spire? Are there 3 or 4 orange bands on a pale horn background?

Yes →

i

No →

j

No

a WINKLE

k

g

l

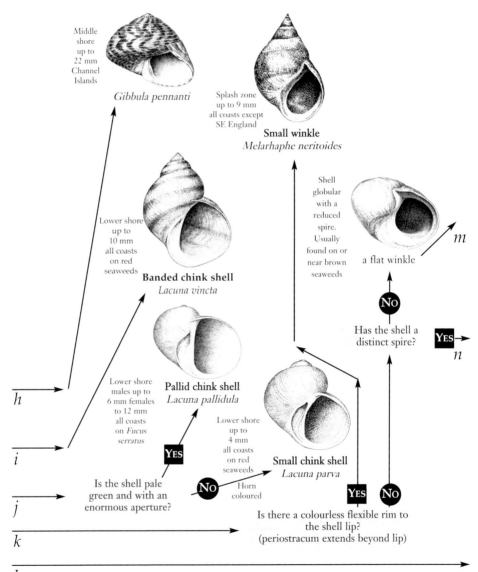

Middle
shore
up to
22 mm
Channel
Islands

Gibbula pennanti

Splash zone
up to 9 mm
all coasts except
SE England

Small winkle
Melarhaphe neritoides

Lower shore
up to
10 mm
all coasts
on red
seaweeds

Banded chink shell
Lacuna vincta

Shell
globular
with a
reduced
spire.
Usually
found on or
near brown
seaweeds

a flat winkle

m

No

Has the shell a
distinct spire?

Yes →

n

Lower shore
males up to
6 mm females
to 12 mm
all coasts
on *Fucus
serratus*

Pallid chink shell
Lacuna pallidula

Lower shore
up to
4 mm
all coasts
on red
seaweeds

Small chink shell
Lacuna parva

h

i

j

Yes

Is the shell pale
green and with an
enormous aperture?

No

Horn
coloured

k

Yes   **No**

Is there a colourless flexible rim to
the shell lip?
(periostracum extends beyond lip)

l

o

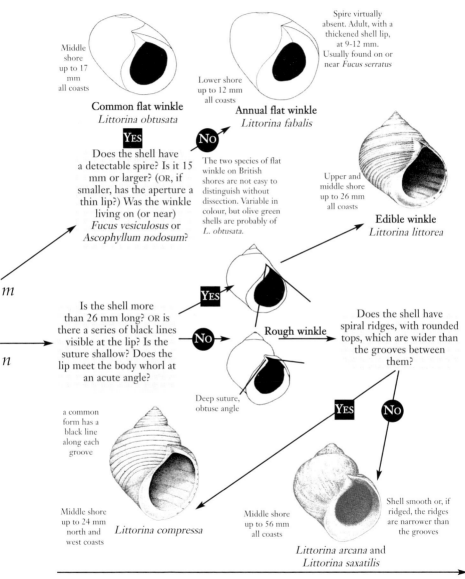

Middle
shore
up to 17
mm
all coasts

Common flat winkle
Littorina obtusata

YES

Does the shell have
a detectable spire? Is it 15
mm or larger? (OR, if
smaller, has the aperture a
thin lip?) Was the winkle
living on (or near)
Fucus vesiculosus or
Ascophyllum nodosum?

Lower shore
up to 12 mm
all coasts

Annual flat winkle
Littorina fabalis

NO

The two species of flat
winkle on British
shores are not easy to
distinguish without
dissection. Variable in
colour, but olive green
shells are probably of
L. obtusata.

Spire virtually
absent. Adult, with a
thickened shell lip,
at 9-12 mm.
Usually found on or
near *Fucus serratus*

Upper and
middle shore
up to 26 mm
all coasts

Edible winkle
Littorina littorea

m

Is the shell more
than 26 mm long? OR is
there a series of black lines
visible at the lip? Is the
suture shallow? Does the
lip meet the body whorl at
an acute angle?

YES

NO

n

Rough winkle

Deep suture,
obtuse angle

Does the shell have
spiral ridges, with rounded
tops, which are wider than
the grooves between
them?

a common
form has a
black line
along each
groove

YES

NO

Middle shore
up to 24 mm
north and
west coasts

Littorina compressa

Middle shore
up to 56 mm
all coasts

Shell smooth or, if
ridged, the ridges
are narrower than
the grooves

Littorina arcana and
Littorina saxatilis

o

p

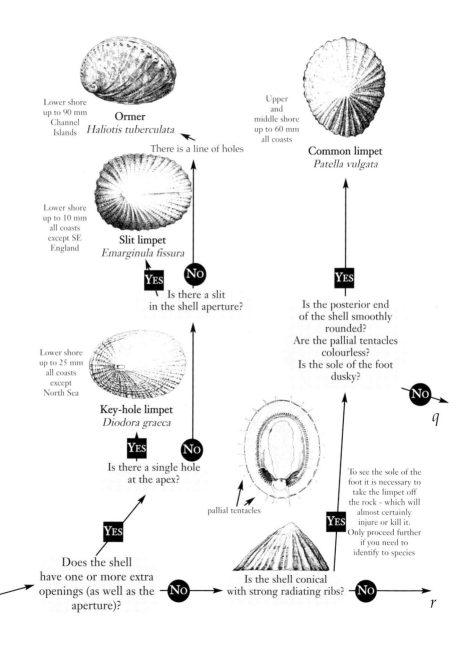

Lower shore up to 90 mm Channel Islands

Ormer
Haliotis tuberculata

There is a line of holes

Upper and middle shore up to 60 mm all coasts

Common limpet
Patella vulgata

Lower shore up to 10 mm all coasts except SE England

Slit limpet
Emarginula fissura

Yes **No**

Is there a slit in the shell aperture?

Yes

Is the posterior end of the shell smoothly rounded?
Are the pallial tentacles colourless?
Is the sole of the foot dusky?

No
q

Lower shore up to 25 mm all coasts except North Sea

Key-hole limpet
Diodora graeca

Yes **No**

Is there a single hole at the apex?

pallial tentacles

To see the sole of the foot it is necessary to take the limpet off the rock - which will almost certainly injure or kill it. Only proceed further if you need to identify to species

Yes

Yes

Does the shell have one or more extra openings (as well as the aperture)?

No

Is the shell conical with strong radiating ribs?

No
r

p

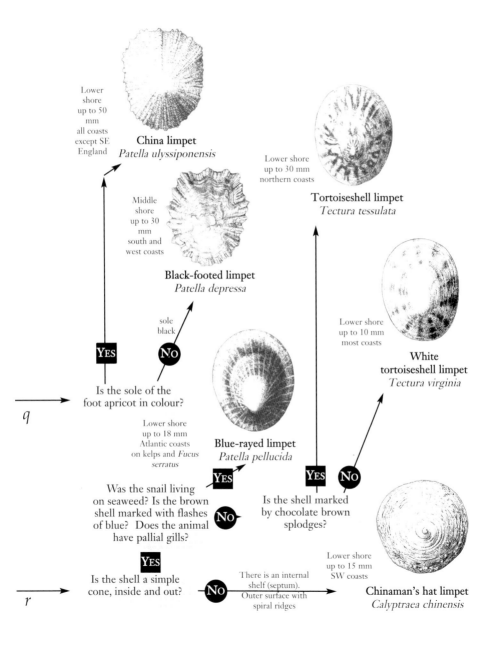

Lower shore up to 50 mm all coasts except SE England

China limpet
Patella ulyssiponensis

Middle shore up to 30 mm south and west coasts

Black-footed limpet
Patella depressa

Lower shore up to 30 mm northern coasts

Tortoiseshell limpet
Tectura tessulata

Lower shore up to 10 mm most coasts

White tortoiseshell limpet
Tectura virginia

sole black

Yes

No

Is the sole of the foot apricot in colour?

q

Lower shore up to 18 mm Atlantic coasts on kelps and *Fucus serratus*

Blue-rayed limpet
Patella pellucida

Yes

Was the snail living on seaweed? Is the brown shell marked with flashes of blue? Does the animal have pallial gills?

No

Yes **No**

Is the shell marked by chocolate brown splodges?

Yes

Is the shell a simple cone, inside and out?

r

No

There is an internal shelf (septum). Outer surface with spiral ridges

Lower shore up to 15 mm SW coasts

Chinaman's hat limpet
Calyptraea chinensis